Innovative Thinking in Risk, Crisis, and Disaster Management

Reviews for
Innovative Thinking in Risk, Crisis, and Disaster Management

This collection is a significant contribution to the literature on risks and disasters in contemporary societies and how to manage them. It is a rich source of new ideas for academics, students, policymakers and practitioners. Written in an accessible style, the book is wide in scope and fertile with ideas. It is timely and challenging and provides new insights and understanding.

John Benyon, University of Leicester, UK and Chair, College of Learned Societies, Academy of Social Sciences

Inspired by discussions over late-modernity, but drawing upon many vivid examples, this collection takes a critical look at contemporary risk management. Issues of the democratization of knowledge, of corporate social responsibility, and of the quality of life powerfully emerge. A book which puts risk firmly in social and political context — with important implications both for academic analysis and institutional practice.

Alan Irwin, Copenhagen Business School, Denmark

Innovative Thinking in Risk, Crisis, and Disaster Management

Edited by
SIMON BENNETT

GOWER

Gower Applied Business Research
Our programme provides leaders, practitioners, scholars and researchers with thought provoking, cutting edge books that combine conceptual insights, interdisciplinary rigour and practical relevance in key areas of business and management.

Published by
Gower Publishing Limited
Wey Court East
Union Road
Farnham
Surrey, GU9 7PT
England

Ashgate Publishing Company
Suite 420
101 Cherry Street
Burlington,
VT 05401-4405
USA

www.gowerpublishing.com

British Library Cataloguing in Publication Data
Innovative thinking in risk, crisis, and disaster
management.
   1. Risk management. 2. Risk perception. 3. Crisis
management. 4. Emergency management.
   I. Bennett, Simon, 1958-
   363.3'47-dc23

Library of Congress Cataloging-in-Publication Data
Innovative thinking in risk, crisis, and disaster management / edited by
Simon Bennett.
      p. cm.
   Includes bibliographical references and index.
   ISBN 978-1-4094-1194-9 (hbk) -- ISBN 978-1-4094-1195-6 (ebk)
   1. Risk--Sociological aspects. 2. Risk assessment. 3. Emergency
management. 4. Crisis management. I. Bennett, Simon, 1958-
   HM1101.I56 2011
   363.1--dc23

                                                                    2011040162

ISBN: 978-1-4094-1194-9 (hbk)
ISBN: 978-1-4094-1195-6 (ebk)

Printed and bound in Great Britain by the
MPG Books Group, UK

Contents

List of Figures

Introduction

Simon Bennett

The theory of late modernity posits the demise of the grand narrative and inception of multiple social realities or world views. In late modernity the old certainties (science as a force for good, for example) are challenged by an increasingly sceptical (or as social theorist Ulrich Beck (1992, 2009) would have it, 'reflexive')[1] public. Respect for venerable authority structures and figures (the church, the debating chamber, the civil service, the medical profession, the legal system, the Fourth Estate, clergymen/women, politicians, civil servants, scientists, doctors, architects, engineers, prosecutors, judges, police officers, etc.) wanes. Respect for anti-establishment figures (like automotive safety activist Ralph Nader, author of the seminal *Unsafe At Any Speed*) and counter-cultures (like the environmental movement) grows. Where modernity celebrated massification and sameness (mass production, mass consumption, mass tourism, mass advertising, agglomeration, standardisation, etc.) late modernity celebrates plurality and differentiation (in tastes, products, lifestyles, beliefs, world views, fashions, etc.) (Hall and Jacques 1989, Bruce 1999, Bauman 2002). At the heart of late modernity lies a paradox: the more science develops and spreads, the more it is questioned:

> *Over the past two centuries the judgment of scientists has replaced tradition in Western societies. Paradoxically, however, the more science and technology permeate and transform life on a global scale, the less this expert authority is taken as a given. In discourses concerning risk, in which questions of normative (self-) limitation also arise, the mass media, parliaments, social movements, governments,*

---

1   For 'reflexive' read reflective.

> *philosophers, lawyers, writers, etc., are winning the right to a say in*
> *decisions. (Beck 2009)*

Education, that venerable motor of heightened consciousness and change, encourages *methodical scepticism*. The Information Society underwrites critique (with data and ideas, conveyed by Wikipedia, Wikileaks and other online repositories). Identities are no longer reflections of pre-existing forms. Rather they are 'works in progress':

> *[W]hat began to emerge in the late 20th Century was a radical shift in*
> *the locus of meaning in western societies from a culture where meaning*
> *and identity were grounded in loyalty to institutions and structures*
> *[like the State and church] to one in which meaning and identity are*
> *grounded in the self as the primary agent of meaning. Overnight the*
> *institutions and structures of the 20th Century quickly entered a*
> *place where their legitimacy was questioned and most loyalty to them*
> *removed. (Roxburgh 2010)*

According to Beck (1992, 2009) there is a suspicion that the 'old' sources of wealth and comfort (like large-scale industry and the development and propagation of new chemical compounds) may pose a threat to human survival (witness concerns about global warming and the synergistic and antagonistic effects of new compounds). Beck calls this dysfunctional aspect of industrial capitalism the boomerang effect. In the 1960s most believed that science offered a solution. That belief has evaporated:

> *[W]e live in a world that has to make decisions concerning its future*
> *under the conditions of manufactured, self-inflicted insecurity. Among*
> *other things, the world can no longer control the dangers produced by*
> *modernity; to be more precise, the belief that modern society can control*
> *the dangers that it itself produces is collapsing – not because of its*
> *omissions and defeats, but because of its triumphs. (Beck 2009)*

Bauman (2002) posits that citizens' concern with 'body-cultivation' heightens awareness of manufactured risks:

> *In view of the centrality of body-cultivation in the activity of self-*
> *constitution, the damage most feared is one that can result in poisoning*
> *or maiming the body through penetration or contact with the skin (the*
> *most massive panics have focussed recently on incidents like mad cow*

disease, listeria in eggs, shrimps fed on poisonous algae, dumping of
toxic waste – with the intensity of fear correlated to the importance
of the body among the self-constituting concerns, rather than to the
statistical significance of the event and extent of the damage).[2]

While few would argue that science has no role to play in solving the world's problems, science as a 'way of knowing' has lost its lustre. Science is facing a crisis of credibility. The problem, says Beck, lies in scientists' inability or unwillingness to admit to the possibility of being wrong. In a time of questioning and scepticism (Beck's reflexive modernity), scientists' unwavering certainty looks out of place, disjunctive, arrogant. By ignoring the zeitgeist – by rejecting openness and participation in favour of secrecy and exclusion – scientists are undermining their own knowledge system. Science is eating itself.

The solution to this crisis of credibility, claims Beck, lies in reflexive scientisation. Where 'unreflexive' science (although 'subject to challenging questions and internal scepticism') is 'dogmatised externally', 'reflexive' science makes explicit its assumptions and precommitments, thereby engaging (or at least not repelling) a sceptical (reflexive) public (Beck 1992).

According to Beck (2009) it is not only science that faces a crisis of credibility but also the forces of law and order. Unable to deal satisfactorily with suicide bombers, hijacked aircraft used as flying bombs and other manifestations of asymmetrical warfare, the State's legitimacy is undermined and victimised populations unnerved:

> *[T]he institutions of prevention, calculability and control are*
> *circumvented. With still relatively 'low' numbers of victims and deeds,*
> *the felt violence and felt war are maximised and explode in the centres*
> *of the felt peace, both literally and in the mass media … How can the*
> *authorities responsible for control and prevention justify their existence*
> *when what they were supposed to prevent is destined to become even*

---

2   Bauman's argument seems to be validated by the reaction of the Japanese and other publics to emissions from the breached Fukushima nuclear plant. *The Guardian*'s Eric Augenbraun (2011), reviewing the Obama administration's advocacy of nuclear power as an environmentally-benign source of baseload, commented: 'While early signs indicate that President Obama's support for nuclear energy will endure the crisis in Japan, it is much more likely that public opinion will not'. Ground was broken on all of North America's 104 nuclear plants before the Three Mile Island accident. Several countries, including Germany and Britain, initiated post-Fukushima safety reviews. Tony Blair and Gordon Brown re-committed Britain to nuclear power (Bennett 2010, 2011). As of 2012 Britain remained committed (Harvey, Vidal and Edwards 2011).

less preventable in the future and threatens to render them superfluous?
(Beck 2009)

In late modernity established loci of authority and power – science, the State – are challenged and, in the case of the forces of law and order, undermined. According to Beck (2009) attempts to resecure the State undermine it further:

> *The restless search for ... lost security begins through measures and strategies that lend the appearance of control and security instead of guaranteeing them and exacerbate the general feeling of insecurity and endangerment ... it is not the terrorist act that destroys the west, but the reaction to its anticipation. It ignites the felt war in the minds and centres of the West.*

This edited collection explores risk assessment and management in the context of Beck's *Risk Society* thesis, and discusses what contribution late modern approaches to risk assessment and management might make to restoring the public's faith in science and technology, in the State's ability to deliver peace and security and to people's quality of life.

The book asks whether a wider ownership of, and participation in, the scientific project (specifically processes of risk assessment and management) would help re-legitimise science. It asks whether multiple social realities can be accommodated within the scientific project to produce a hybrid, more socially-acceptable *way of knowing and acting*. It evaluates such pro-participation discourses as Irwin's (1995) *Technological Citizenship* and Zimmerman's (1995) *Technological Governance*. Fundamentally the book explores novel approaches to managing risk, crisis and disaster at a time when risk – social, economic, political and environmental – looms large in private and public discourse (Hutton 1995, Bauman 2002, Beck 1992, 2009, Cable 2009).[3]

At several points the book touches on economic risk as a source of public disquiet. Not only are the public subjected to sudden shocks (the financial crisis that began in 2008, for example, and consequent austerity measures), but also to the long-term corporate trend of 'hollowing-out':

> *Firms, anxious to meet their shareholders' requirements for high returns, have taken advantage of the deregulation of the labour market*

---

3   The book reviews the management of risks pertaining to numerous domains – science, technology, informatics, work, organisation, public safety and the natural environment.

to impose more casualised, part-time, 'flexible' patterns of work and so increase their capacity to modify their costs to the changing pattern of demand. Only around 40 per cent of the work-force enjoy tenured, full-time employment or secure self-employment ... another 30 per cent are insecurely self-employed, involuntarily part time or casual workers; while the bottom 30 per cent, the marginalised, are idle or working for poverty wages ... the wider economic and social impact has been disastrous. (Hutton 1995)

A number of short case studies are presented below by way of scene setting. The first case study outlines the role played by local residents (non-scientists) in risk-managing the 1989 Exxon Valdez oil spill. The second describes how patient groups are helping to improve the design of clinical trials. The third reviews the United Kingdom government's efforts (in the midst of cutbacks) to encourage public participation in social projects. The fourth and fifth describe what can happen when science discounts lay understandings.

The Public and Prince William Sound: A Lesson for Science?

As it pertains to risk assessment and management, late modernity is a *state of mind* that legitimises subjectivity and voluntarism. It encourages rather than rebuffs contributions from 'non-experts'. Following the grounding of the oil tanker Exxon Valdez at Prince William Sound, the responsible authorities (government, coast guard, oil company, etc.) activated their contingency plans. The results were less than satisfactory. There seemed to be little to lose from listening to the public (fishermen, businesspeople, community activists, etc.). No attempt was made to reconcile the differing world views. Regarding how best to manage the disaster and prevent future spills, dissonance was seen as a positive; a creative rather than a destructive process; a safeguard against group-think and hegemony:

Having complementary and countering input from many views and voices as an integral part of planning, prevention and response can productively exploit the problematics of ... multiple realities in ways which would not be possible in a plan which attempted to reconcile them ... The apparent chaos of many viewpoints and many voices allows a larger view of order to emerge, one without the disqualifying filters of the self-delusional, intentional bias that allows systems to lurch unforeseen and unintended toward normal accidents. (Browning and Shetler 1992)

Late modernity – to the extent that it legitimises and exploits multiple social realities – provides a safeguard against the dictatorship of a single (and potentially flawed) approach to risk assessment and management.

Patients and Clinical Trials

In the United States the Multiple Myeloma Research Foundation (MMRF) influences the conduct of research into multiple myeloma, a cancer of the bone marrow that kills a majority of sufferers within five years. The MMRF uses economic levers (it is able to disburse around $30 million each year) to bring new treatments to market as quickly as possible:

> For researchers … papers and grants, not treatments, are the measure of success, and results are kept secret until publication. [The MMRF], however, wants treatments to emerge as quickly as possible, so one condition of receiving funding … is that results must be shared. The foundation … holds researchers accountable when progress slows. (Giles 2008)

Up to speed with the latest research and with an intimate knowledge of the disease, MMRF patient advocates have been able to influence the conduct of clinical trials:

> Advocates can … provide unique input into treatments. When a group of oncologists designed a study to test the combination of lenalidomide and the steroid dexamethasone, patient advocates objected to the proposed steroid dose as it can make users unpleasantly hyperactive … Reluctantly the oncologists agreed to also trial a lower dose of the steroid – a decision that was vindicated when survival rates in the low-dose group proved to be significantly higher. (Giles 2008)

Here we see how 'lay' knowledge (of an especially intimate nature) can inform – indeed, improve – scientific experimentation. We also see a persistent reluctance to acknowledge lay expertise.

A Politician's Interpretation of the 'Active Citizen' Model

Following the 2010 instalment of a new government in the midst of a financial crisis, Prime Minister David Cameron launched his Big Society initiative – an attempt to ratchet-up community activism. The Prime Minister framed the scheme as a way of both tapping into community spirit and filling the holes left by cuts:

> [L]et me briefly explain what the Big Society is ... The Big Society is about a huge culture change, where people, in their everyday lives, in their homes, in their neighbourhoods, in their workplace don't always turn to officials, local authorities or central government for answers to the problems they face, but instead feel both free and powerful enough to help themselves and their own communities ... For years, there was the basic assumption at the heart of government that the way to improve things in society was to micromanage from the centre, from Westminster. But this just doesn't work. We've got the biggest budget deficit in the G20. And over the past decade, many of our most pressing social problems got worse, not better. It's time for something different, something bold – something that doesn't just pour money down the throat of wasteful, top-down government schemes. (Cameron 2010)

For Cameron, the State is too big and remote to understand the needs of local communities. Better, then, to capitalise on local knowledge, insight and enthusiasm:

> If you've got an idea to make life better, if you want to improve your local area, don't just think about it – tell us what you want to do and we will try and give you the tools to make things happen. I passionately believe what we have begun here will spread right across our country – covering it in innovation, local inspiration and civic action. (Cameron 2010)

The launch of the Big Society project in the midst of a recession begs the question: Is the government convinced that at active citizenry is better than a passive one, or is it cynically jettisoning its responsibilities to save money with little regard for street-level outcomes? Soon-to-be Labour leader Ed Miliband said: 'Cameron's government is cynically attempting to dignify its cuts agenda by dressing up the withdrawal of support with the language of reinvigorating civic society' (Miliband cited in Prince 2010). Could it be that politicians (or businesspeople or bureaucrats or other authority figures) see active citizenship as a way of spreading the political costs of failure (on the basis that the more people you involve, the less opprobrium you attract to yourself if things go wrong)? Or is that too cynical a view of politicians?

The Exxon spill provided a shop window for lay knowledge, the post-disaster clean-up benefitting from community involvement. Helped by the lure of research endowments, the MMRF's patient advocates are able to influence the design of clinical trials. These examples of engagement may be the exception, however. Traditionally scientists and technologists have been reluctant to consult non-experts, as the following two case studies demonstrate.

Chernobyl Fallout

Following the 1986 destruction of the Chernobyl nuclear reactor, a radioactive cloud drifted across northern Europe. In Britain radioactive fallout was deposited on the Cumbrian Hills. Faced with the possibility that radioactivity would enter the food chain via grazing sheep the Ministry of Agriculture, Fisheries and Food (MAFF) despatched scientists to risk-manage the situation. Rather than exploit the accumulated knowledge of Cumbria's hill farmers, MAFF's scientists used a textbook model of soil conditions, animal husbandry and meteorological conditions. MAFF's 'naive' science resulted in mistakes. Scientists' assumptions about grazing were inaccurate, as were their assumptions about upland soil types and substitute feeds. One MAFF scientist told the farmers to feed their sheep on imported straw. The farmers were incredulous. As one put it: 'I've never heard of a sheep that would even look at straw as fodder. When you hear things like that it makes your hair stand on end' (cited in Irwin and Wynne 1996). MAFF's indifference to knowledgeable client groups damaged scientists' public image and undermined their relationship with the community. MAFF's dismissal of farmers' knowledge undermined their identity (Wynne 1996). (The sociological approach to risk assessment and management treats all accounts as equally valid – if not equally relevant.) Barry (2000) comments:

> [A]s Wynne demonstrates, scientists do not have a monopoly on valid knowledge. In his now classic account of Cumbrian hill farmers [Professor Brian Wynne] notes how farmers possessed a knowledge of the Lake district which was simply unavailable to the scientists.

The Cumbrian hill farms episode suggests that highly professionalised and ordered practitioner communities may become self-referential – hubristic, even. It would be interesting to know whether the organising mechanisms and associated behaviours of practitioner communities (like the Royal Colleges) have the effect first, of insulating members from non-members, and secondly, of cutting members off from useful knowledge (residing in other practitioner communities and society at large). Are the Royal Colleges and Societies no more than 'knowledge silos'? When it comes to understanding and managing aetiologically complex real-world risks, hubris gets in the way. Speaking to the hill farms episode, Mooney (2008) comments:

> [T]he sheep farmers became increasingly distrustful of the arrogant assertions of government scientists who didn't seem to credit their own sophisticated understanding of all things sheep-related. And they had every right to be. The scientists weren't communicating or even taking their audience seriously, and so kept making fairly bone-headed mistakes ... Scientists have long held to a kind of classroom-oriented, one-way model for the dissemination of their knowledge – e.g., they know the science, they tell it, the public understands it and accepts it.

To what degree are the public respected, and to what degree patronised by scientists?[4]

Farm Labouring

In the 1980s a bitter dispute erupted between the National Union of Agricultural and Allied Workers (NUAAW) and the government's Advisory Committee on Pesticides (ACP). Drawing on its research into the use of the pesticide 2,4,5-T (a defoliant with potential long-term adverse health effects) the NUAAW argued that it should be banned (Irwin 1995). The pesticide had already been proscribed in Canada, the United States and the former Soviet Union. In Britain it had been dropped by British Rail and the National Coal Board.

Drawing on a de-contextualised, laboratory-based model of farmworking the ACP argued that 2,4,5-T was safe if used 'in the recommended way and for the recommended purposes'. The farmworkers argued that real-world constraints and conditions made it difficult – if not impossible – for farm labourers to follow the ACP's instructions for use. The consequence? The chemical might be used in an unsafe manner, exposing farm labourers (and their families) to the risk of contamination.

Drawing on farm labourers' experience the NUAAW argued that in a time-pressured environment labourers might have no choice but to spray in

4   Ironically, scientists' salaries are often funded from the public purse.

windy conditions; that protective clothing was not always available; that in hot and humid conditions labourers might choose not to wear protective clothing anyway; that under rough and ready field conditions diluting the chemical with the required degree of accuracy was problematic; that time pressures and logistical difficulties meant that soiled clothing might be worn home (exposing family members to the risk of chemical contamination); that time pressures and logistical difficulties meant that spraying equipment might not be cleaned; and finally, that farm labourers' dependence on an employer's beneficence might lead them to ignore the safety advice (to keep the farmer 'on-side'). It is difficult for employees to play it by the book in isolated communities with few employment opportunities. Accommodating employers' agendas is the lived reality of life in a rural community.

The debate over 2,4,5-T is noteworthy for the way in which it polarised opinion. The ACP dismissed the NUAAW's data as unscientific. The NUAAW dismissed the ACP's guidelines as unworkable. Concern about the lack of engagement between scientists and client groups has boosted interest in risk communication theory.

Risk Communication

Risk communication represents a *sociological* approach to the study of risk assessment and management. Risk communication theorists concern themselves with the dialogue, or rather the lack of dialogue, between experts and lay people (Hadden 1989). The crux of Wynne's work is a critique of the distinction which has been drawn between expert and lay decision-makers in much of the psychological and sociological work on risk. Wynne argues that the expert approach is based on a misconception of science – specifically of an objective scientific community in which subjective factors (social, cultural and psychological) either do not, or at least minimally, influence the decision-making process.

Risk communication theorists argue that lay people's perceptions have been tied to a particular set of social, cultural and psychological factors. Lay perceptions are frequently characterised as being constructed on the basis of irrational and non-objective models of reality, which are then validated with reference to folk theories of risk and danger. Sociologists concerned with risk communication concentrate on the transfer of information between experts and lay people. They argue that experts typically count lost lives (the quantitative approach to risk assessment and management) while the general public focuses on other factors, in particular fairness and controllability (the qualitative approach to risk assessment and management). Crucially, they argue that lay people may possess useful knowledge. As one farm labourer (cited in Irwin 1995) who regularly sprayed 2,4,5-T put it, with not a little incredulity:

> They [the 'experts'] may know the risks of 2,4,5-T. They may handle the stuff properly. They tell us we'll be alright if we use the spray normally. But have they any idea what 'normally' means in the fields?

The Participation Argument

Few democrats would argue that widening public participation in decision-making is a bad thing. As mentioned above one of the defining characteristics of late modernity and of the Risk Society is a general loss of faith in the scientific project and in experts' ability to control the direction and products of scientific inquiry. Unfortunately for the advocates of active citizenship another characteristic of late modernity is a general loss of interest in formal participatory mechanisms like general elections. It is one thing to espouse active citizenship but quite another to persuade citizens to become active. There is a gap between (idealistic) rhetoric and reality:

> It has become a commonplace in both contemporary political philosophy and contemporary politics to argue that active citizenship is a good thing ... Broadly speaking one can see that the revitalisation of the ideal of active citizenship developing in the wake of decline of traditional welfarist and social democratic notions of politics. Today, these are often regarded as implying a too passive model of civic engagement. I don't want to discuss these arguments about active citizenship in detail except to note that what is forgotten in these discussions – which are often rather general and normative – is the problem of how such virtuous persons come about. In her book The Will to Empower Barbara Cruikshank notes that 'in reformist and democratic discourses, citizenship and self-government are tirelessly put forward as solutions to poverty, political apathy, powerlessness, crime, and innumerable other problems' ... But, as she notes, the value placed on active citizenship raises a whole series of questions: How are active citizens made? In what circumstances is it possible for a person to be active?

> *And how is 'activity' performed? In its concern with establishing some general account of the need and the justification for 'active citizenship' political theory fails to think about all the practical, psychological and physical problems of becoming active. (Barry 2000)*

If active citizenship is to contribute to the more effective management of the multiplying hazards of the Risk Society, ways have to be found to facilitate involvement.

A second problem lies with the value of lay knowledge. As Barry (2000) puts it:

> *Although some non-experts have valuable knowledge we cannot assume that all non-experts have knowledge of equal value. Many non-experts are likely to be uninformed or have little to contribute.*

A third difficulty concerns the issue of time commitment (getting involved eats into leisure time):

> *To become an active citizen — to take part in a debate, or a form of deliberative politics, engage in a form of direct action, to disagree — is a performance. And it is a performance which requires and has costs — not just in time and loss of earnings. But in terms of exposure, personal relationships. (Barry 2000)*

Lastly there is the question of representativeness — it is not unknown for special interest groups to be hijacked by a cliqué. Members of Parliament and councillors can be voted out. This sanction may not apply to self-appointed community 'leaders'.

Conclusion

Late modern societies are reflexive societies. The questions (but not necessarily the answers) multiply. In late-modernity the things that used to make us feel secure (science, technology, medicine, the State, the judiciary, the law, the banks, bureaucracy, well-intentioned political projects and other facets of modernity) unnerve us (witness the impact of the 2008–2012 global financial crisis on public confidence and politicians' and citizens' changing perceptions of the EU blueprint for 'ever-closer union').[5] In the Risk Society civic action has less to do with advancing class interests than with mitigating the effects of techno-scientific hazards (like global warming). For some the Risk Society is an opportunity to create a more democratic and inclusive social order where citizens' opinions are valued, their ideas recognised and their talents utilised (Browning and Shetler 1992). Irwin (2001) calls for 'the development of an open and critical discussion between researchers, policy makers and citizens'. Rabeharisoa and Callon (1999) argue for what they term *cooperative research* between medical researchers and patients, where the latter contribute to the production of scientific knowledge. Perrow (1984), concerned about the propensity of tightly coupled complex systems to succumb to unanticipated interactions, makes a case for the interrogation of mental models within and across organisational strata.[6] For Weick and Sutcliffe (2001) 'mindfulness' – the questioning of assumptions and methods and exploitation of employee insights – helps improve organisational reliability. This book explores how risk, crisis and disaster management might work in a Risk Society and how those responsible for public well-being might usefully exploit the late-modern zeitgeist. Fundamentally the book is a vehicle for 'out-of-the-box' thinking.[7]

---

5   By the end of 2011 it was clear that Britain and Germany framed the EU project in quite different ways. While Prime Minister Cameron was happy to use Britain's veto to (ostensibly) defend the City of London from 'unnecessary' regulation (it is worth remembering that the global financial crisis that began in 2008 was triggered by banks' irresponsible sub-prime lending), the Germans applied a somewhat larger (and less parochial?) frame of reference. As Germany's foreign minister remarked during a post-veto bridge-repairing visit to London: 'For us, Europe is much more than a currency or a single market. It is a key question ... it is a political union that we want' (Westerwelle cited in Attewill 2011).

6   Crew resource management (CRM), which has helped improve aviation safety, encourages flight and cabin-crew to monitor and cross-check colleagues' assumptions, decisions and actions. It creates an inquisitorial mind-set (Krause 1996).

7   The Cameron–Clegg government's efforts to re-balance the British economy may be considered an example of out-of-the-box thinking. Ever since the Thatcher-Regan era the received wisdom has been that Britain's industrial sector is doomed – the inevitable victim of non-unionised, sweated labour in developing economies like China, India and Brazil (Chu 2011). Not so, claims the Con-Lib coalition, who talk of re-balancing a British economy that has become ever-more dependent on tax revenue from the City of London which, as the toxic debt crisis demonstrated, is prone to poor decision-making. The Con-Libs may be on to something. As wage levels in the BRIC (Brazil, Russia, India and China) countries rise, capital will look to relocate. A British economy that is pro-industry would be in a position to attract unanchored capital. Entrepreneur James Dyson made the case for re-balancing in his 2010 report *Ingenious Britain*: 'Now, more than at any time over the past twenty years, I sense there is a real opportunity to set a new vision for our economy. To do this, a new government must take immediate action to put science and engineering at the centre of its thinking – in business, industry, education and, crucially, in public culture' (Dyson 2010).

Chapters, Arguments and Authors

EMPOWERING EMERGENCY RESPONDERS, ROGER MILES

Risk management and emergency response practices in the United Kingdom have undergone major change. New legal and professional best-practice frameworks emphasise the need for integration. In his discourse on the Risk Society Beck cautions against over-confidence in the efficacy of risk management and crisis response measures. He queries whether formal systems can deal effectively with the unexpected. Two observations of contemporary practice inform this chapter. First, although training exercises are vital for establishing preparedness, the scenarios used in such exercises often emphasise the use of formal command and control procedures rather than promotion of personal effectiveness and leadership – especially at the lower levels. Secondly, reports on live incidents rarely highlight where initiatives at lower levels improved matters, despite there being anecdotal evidence of the effectiveness of individual enterprise. The author makes the case for enabling action. Examples are provided of successful enabling environments. Lessons are drawn from high-reliability organisations (like resilient business-supply chains). Research into how individual performances in demanding contexts can be enhanced is discussed.

Roger Miles has an MPhil in Occupational Psychology and is a Fellow of the British Psychological Society (BPS). Roger ended his Civil Service career as Head of the Emergency Planning Division at the Home Office. His consultancy work has included evaluations of contingency planning for local authorities, Railtrack, the Environment Agency and the Welsh Office. His activities for the BPS have included chairing the Board of Examiners for Occupational Psychology. He has served as a Reviewer for the Quality Assurance Agency for Higher Education and as External Examiner for the Hull University MSc in Occupational Psychology.

TERRORISM AND THE RISK SOCIETY, DAVID WADDINGTON AND KERRY MCSEVENY

This chapter considers the usefulness of the application of Ulrich Beck's Risk Society approach to contemporary 'post-9/11' terrorism. Beck argues that the events of 11 September 2001 have drastically altered common understandings and perceptions of terrorism, in ways which make it comparable to other global manufactured risks such as climate change or financial crisis. This 'new terrorism' is discursively constructed by the State and mass media as an unpredictable and de-territorialised risk of potential attack.

Beck's Risk Society theory provides convincing evidence of the relationship between globalisation and the seemingly borderless nature of terrorist activity, but his interpretation has been subject to criticism for its misrepresentation of the scope and scale of terrorism (e.g. Mythen and Walklate 2008, Stohl 2008). In addition to an overview of these criticisms, the chapter provides a more detailed exploration of two aspects of the response to terrorist acts (namely the political manipulation of representations of terrorist activity and subsequent policy decisions in relation to national security) which, although briefly and superficially addressed by Beck, are significant enough to demand further attention. Finally, the chapter addresses a key omission in Beck's work by outlining recent attempts to engage with the underlying motivation for acts of terrorism, and considers the implications for current foreign and domestic policies.

Kerry McSeveny, a research assistant at Sheffield Hallam University, has recently completed a PhD on the interactional management of identity and accountability. Kerry's research interests include communication and conflict, language, gender and power, neoliberal approaches to health and the representation of social issues. Recent projects have investigated public confidence in local policing and the use of social media in managing protest situations.

David Waddington is Professor of Communications and Head of the Communications and Computing Research Centre at Sheffield Hallam University. His main research interest is the policing of social and political conflict. He is author of *Contemporary Issues in Public Disorder: A Comparative and Historical Approach* and *Policing Public Disorder: Theory and Practice*.

THE EMERGENT NATURE OF RISK AS A PRODUCT OF 'HETEROGENEOUS ENGINEERING': A RELATIONAL ANALYSIS OF OIL AND GAS INDUSTRY SAFETY CULTURE, ANTHONY J. MASYS

As systems become larger, more complex and more interdependent, new methods are required to identify latent risks. Events such as the 1988 Occidental Piper Alpha accident, the 2003 Shell Brent Bravo accident and the 2010 British Petroleum Deepwater Horizon Oil Rig disaster in the Gulf of Mexico highlight the risks associated with operating complex socio-technical systems. Industry is showing an increased interest in the concept of 'safety culture' as a way of reducing the potential for large-scale accidents associated with routine tasks (Cooper 2000). Perrow argues that the sharing and questioning of mental models and world views across strata can help improve safety. Late-modern approaches to risk, crisis and disaster management both recognise and exploit complexity, specifically the existence of different world views (reality constructs). This chapter uses Actor Network Theory (ANT) to deconstruct, through a relational analysis, the safety culture of the oil and gas industry. In congruence with the late-modern approach to risk management, ANT, with its inherent multi-vocality, shows that safety culture emerges from a network of heterogeneous, interdependent elements. The author argues that if safety is to be improved there must be a broadening of participation among stakeholders, with specific consideration of the interdependencies between the physical, human and informational domains.

Anthony J. Masys has a PhD in the management of socio-technical risk from the University of Leicester. He also has a BSc in Physics and an MSc in Underwater Acoustics and Oceanography. Dr Masys works as a Defence Scientist at Defence Research and Development Canada's (DRDC's) Centre for Security Science (CSS). A former Royal Canadian Air Force (RCAF) officer and naval aviator, he has flown operationally in support of crisis and disaster management efforts. He was involved in the search and recovery effort that followed the on-board fire and ditching of Swissair Flight 111 off the Canadian coast.

THE INHUMAN: RISK AND THE SOCIAL IMPACT OF INFORMATION AND COMMUNICATION TECHNOLOGIES, DAVID ALFORD

This chapter considers the possible dangers associated with the rapid growth and permeation of information and communication technologies (ICTs). If the unfettered development of ICTs is risky we must address the matter of regulation. Ulrich Beck describes how the negative effects of new technologies rebound upon society at large (the boomerang effect). He also argues that late modernity is characterised by a diffusion of responsibility ('organised irresponsibility'). These issues are highly relevant to computer development. Lyotard framed the potential problem of ICTs in discussing the nature of knowledge in computerised societies, and how the encroachment of the inhuman subordinates the human. We are entering a new phase of potent technological convergence. Computing devices permeate every dimension of human activity. Worryingly, in the absence of boundaries and ethical guidelines, we lack the means to evaluate the consequences. Today our technological scope far exceeds our predictive ability. Late-modern approaches to risk assessment and management recognise the potential benefits of public participation – though this is undoubtedly a problematic area. The chapter draws on a wide range of literature in its consideration of: the dynamics and consequences of ICT expansion; computer evolution and potential social concerns; approaches to computer ethics; democratisation of knowledge. The text addresses social, technical, organisational and educational concerns and attempts to blend aspects of post-modern and late-modern discourses. Central to the long-term success of ICT regulation is a shift in attitudes, both in the ethical stance of precaution and the democratic process of public involvement.

David Alford has a BSc and MPhil in Psychology and a PhD in cognitive psychology. He has held numerous academic posts. Between 2003–05 David was a leading member of the EPSRC's Risk Perception and Assessment in Design Network, an initiative to better align the worlds of academia, business and industry. He has contributed to publications in the fields of experimental psychology, cognitive science, acoustics, mechanical engineering, thinking and reasoning, and computing.

RISK AS WORKERS' REMEMBERED UTILITY IN THE LATE-MODERN ECONOMY, CLIVE SMALLMAN AND ANDREW M. ROBINSON

The Risk Society thesis encapsulates a number of dilemmas, including the future of employment practices in a world where scarcity and perturbations (social, economic, political and environmental) are normal. Robinson and Smallman evaluate the risks posed to employees by workplace opportunities and threats in the late-modern economy. Workplace risk has important consequences for the management of enterprise, especially in regard to three significant themes: the nature of jobs ('good' versus 'bad' (Coats 2006, Doeringer and Piore 1971)); the form of employment arrangements (Kunda, Barley and Evans 2002, Smith 2001); and the growth of high-performance workplaces (Appelbaum, Bailey and Berg 2000, Whitfield 2000) organised around flexible production (Hirst and Zeitlin 1991, Marshall 1999, Piore and Sabel 1984, Sabel 1989).

Clive Smallman has been Professor of Business Management and Head of Commerce Research at Lincoln University, New Zealand, since 2003. Recently appointed a Fellow of the New Zealand Work and Labour Market Institute (Auckland University of Technology), he was formerly the James Tye/British Safety Council Fellow at the Judge Business School, Cambridge.

Andrew M. Robinson is a Senior Lecturer in Accounting and Finance at the University of Leeds's Centre for Employment Relations, Innovation and Change. Andrew conducts research in the fields of profit-sharing, employee share ownership and governance, health and safety at work and the economics of information and organisations. Much of his work aims to develop a better understanding of the nature of firm-level performance and the factors that contribute to it. Primarily this focuses on the role and behaviour of employees in this process and whether 'participative' or 'co-operative' forms of governance are compatible with company profitability and shareholder financial success.

AVIATION AND CORPORATE SOCIAL RESPONSIBILITY, SIMON BENNETT

In recent years the cost of air travel has declined dramatically in real terms. The public has developed a taste for cheap air travel. This taste is satisfied by an intensely competitive global aviation industry in which cost-cutting is *de rigueur*. The global financial crisis and escalating fuel costs have encouraged airlines to bear down on employment costs (Massachusetts Institute of Technology 2011). Pilots' terms and conditions have been eroded. Salaries, especially those offered to new entrants, have fallen. Unable to afford accommodation close to major airports pilots may commute long distances, with potential safety impacts (long-distance commuting can cause fatigue and stress in flight crew). It would appear that, at least in regard to aviation, society is more concerned about cost than risk. This consumerist calculation contradicts Beck's thesis that, in the Risk Society, attention shifts from economic to existential questions. Society's apparent lack of concern about the impact of cost-cutting on operational risk (and, for that matter, the impact of aviation on the environment) is a contra-indication. Is Beck's Risk Society thesis sufficiently elastic to accommodate the choices made by consumers of air travel?[8] An 'end-of-pipe' solution to the problem of commuting is offered – subsidised accommodation for low-paid flight crew.

Simon Bennett is Director of the University of Leicester's Civil Safety and Security Unit (CSSU). He has a BA in Public Administration, an MSc in Communications and Technology and a PhD in the Sociology of Scientific Knowledge (SSK) from Brunel University, Middlesex. His research interests include aviation human factors and the management of socio-technical risk. His books include *Londonland: An Ethnography of Labour in a World City*, *Human Error – By Design?* and *A Sociology of Commercial Flight Crew*. In 2010/2011

---

8   While minor shifts and reformulations can shore-up a theory for a while, in time the weight of contra-indications will cause it to collapse. A paradigm shift ensues.

he worked for the British Air Line Pilots' Association (BALPA) where he investigated the lived reality of commercial air operations. The resulting 222-page report *The Pilot Lifestyle – A Sociological Study of the Commercial Pilot's Work and Home Life* can be obtained from the Institute of Lifelong Learning at the University of Leicester.

INVESTIGATING RESILIENCE, THROUGH 'BEFORE AND AFTER' PERSPECTIVES ON RESIDUAL RISK, HUGH DEEMING, REBECCA WHITTLE AND WILL MEDD

This chapter argues that the recent shift towards the Flood Risk Management (FRM) approach, which has shifted responsibility onto the individual, is an example of the Risk Society at work. Decades of support for structural solutions, combined with the increasing challenges of climate change, have seen the expansion of communities into flood-prone areas. The research findings presented here (derived from 'before' and 'after' studies of flood-affected communities) show how the government's policy of 'Making Space for Water' (Department for Food and Rural Affairs 2005) plays out. In a clear reference to sociological understandings of risk communication (Hadden 1989, Irwin 1995) Deeming, Whittle and Medd argue for citizens to be more involved in the decisions that are made around flood risk management, and for better support for the process of flood recovery.

Hugh Deeming is a Senior Research Associate at both Lancaster University and the United Kingdom Emergency Planning College, Easingwold. An ethnographer, he investigates the human aspects of natural disasters. He has contributed to our understanding of Britain's recent flood events, including those at Hull, Sheffield and Doncaster.

Rebecca Whittle (née Sims) is a Senior Research Associate in The Lancaster Environment Centre, Lancaster University. She is interested in the sustainability of community–environment relations and has spent the last four years researching natural disasters from a social-science perspective. She researches long-term disaster recovery and risk governance.

Will Medd is Lecturer in Human Geography at Lancaster University. His research interests include the relationship between everyday practice and the governance of socio-technical systems. He has investigated water demand, drought and flood.

MANAGING RISKS IN A CLIMATICALLY DYNAMIC ENVIRONMENT: HOW GLOBAL CLIMATE CHANGE PRESENTS RISKS, CHALLENGES AND OPPORTUNITIES, TODD HIGGINS

Theories of risk, crisis and disaster management are discussed in the context of global climate change, with an emphasis on plants and food crops. The Risk Society may have shifted discussion from the effects that climate change will have on human and other living populations to how humans have visited climate change upon the earth. Possibilities for mitigating the effects of global climate change through the adaptation of existing technology and application of focused risk management are discussed. Risk management strategies that allow populations to take advantage of opportunities presented by global climate change are reviewed. The chapter does not discuss the validity of the science of global climate change: it is argued that the climate change process predates industrialisation. The view is put that the greatest advantage we have as we experience these events is our capacity to prepare for, and adapt to, global climate change.

Todd Higgins has a BS degree in Plant and Soil Science and a PhD in agronomy. Todd served in the United States Army and Missouri National Guard. From 2002 to 2005 he was the deputy director, then director of the Military Support to Civil Authorities directorate of the Missouri National Guard. Todd has completed the US Federal Emergency Management Agency's (FEMA's) Emergency Management Institute's Master Exercise Practitioner Programme. In 2007 he was appointed Director of the Land Mine Detection Research Center at Lincoln University, Missouri.

A FUTURE FOR LATE-MODERN SOCIAL FORMATIONS IN DETROIT?, SIMON BENNETT

In recent years Detroit, America's Motor City, has been in decline. The city's economic slump has had profound social consequences. Wealthier Detroiters have moved to the suburbs, leaving the city centre populated by the disadvantaged. Whole neighbourhoods have been razed or left to decay. Social commentator Rebecca Solnit (2007) notes: 'Some forty square miles has evolved past decrepitude into vacancy and prairie'. The economic vacuum left by fleeing capital has been partly filled by a 'bottom-up' urban agriculture movement. In typical late-modern fashion native Detroiters, supported by curious newcomers, have begun to plant Detroit's myriad vacant lots. Detroit's spontaneous urban agriculture movement has had economic and psychological benefits: able to feed themselves, city dwellers have regained their self-confidence. Predictably, perhaps, Detroit's urban agriculture movement faces the prospect of being swallowed up by agribusiness. Land is cheap. Venture capitalists are looking for money-spinning projects. There is a danger that, in time, agribusiness may consume Detroit's small-scale growers. Are we about to witness the resurrection of the Modernist project in Detroit City? If big business does annihilate Detroit's innovative small-scale entrepreneurs, what does this say about the durability of late-modern social formations? What does it say about theories of late-modernity? The Detroit experience suggests that far from shifting, epochs splinter and intermingle, creating a kaleidoscope of social and organisational forms, some well resourced (in terms of money, political patronage and influence), others not.

All those who contributed to this book are connected in some way to the University of Leicester's Civil Safety and Security Unit (CSSU).[9] The Unit delivers two Masters programmes (in Risk, Crisis and Disaster Management and Emergency Planning Management) and has over 270 distance-learning students worldwide. The Unit, which has two full-time academics and two full-time administrators, was spun out of the Department of Criminology (formerly the Scarman Centre). Most of the authors are Associate Tutors to the two Masters programmes.[10] CSSU academics supervise MPhil and PhD students and consult to government, industry and business. Projects have been undertaken for the Cabinet Office (national emergency preparedness), an engineering consultancy (a critique of railway safety rules) and several airlines (including easyJet and DHL Air). PhD topics have included the prevention of blue-on-blue incidents (fratricide) amongst NATO air forces and the migration of teambuilding techniques between commercial aviation and healthcare.

References

Appelbaum, E., Bailey, T. and Berg, P. 2000. *Manufacturing Advantage: Why High Performance Work Systems Pay Off*. London: Cornell University Press.
Attewill, F. 2011. Osborne: EU won't get £30bn out of us. *Metro*, 20 December.
Augenbraun, E. 2011. Fukushima fears chill Obama's atomic ambition. The president is a big supporter of the US nuclear industry – and it's mutual. But public opinion after Japan's disaster is otherwise. *The Guardian*, 23 March.

---

9   Headquartered at 14 Salisbury Road, Leicester, LE1 7QR.
10 Associate Tutors write teaching material, supervise research, lecture at Study Schools and mark essays and dissertations. They are *de facto* academics. Most CSSU Associate Tutors have been associated with the Unit for over ten years.

Barry, A. 2000. Making the Active Scientific Citizen. Proceedings of the 4S/ EASST conference, 'Technoscience, Citizenship and Culture', University of Vienna, 28–30 September, 2000.

Bauman, Z. 2002. A sociological theory of postmodernity, in *Contemporary Sociological Theory*, edited by C. Calhoun, J. Gerteis, J. Moody, S. Pfaff and I. Virk. Oxford: Blackwell, 429–40.

Beck, U. 1992. *Risk Society: Towards a New Modernity*. London: Sage.

Beck, U. 2009. *World at Risk*. Cambridge: Polity.

Bennett, S.A. 2010. *Insecurity in the Supply of Electrical Energy: An Emerging Threat to Information and Communication Technologies?* Faringdon: Libri Publishing Ltd.

Bennett, S.A. 2011. Insecurity in the supply of electrical energy: An emerging threat? *The Electricity Journal*, 24(10), 51–69.

Bennett, S.A. 2011. *The Pilot Lifestyle: A Sociological Study of the Commercial Pilot's Work and Home Life*. Leicester: Vaughan College, University of Leicester.

Browning, L.D. and Shetler, J.C. 1992. Communication in crisis, communication in recovery: A postmodern commentary on the Exxon Valdez disaster. *International Journal of Mass Emergencies and Disasters*, 10(3), 477–98.

Bruce, S. 1999. *Sociology: A Very Short Introduction*. Oxford: Oxford University Press.

Cable, V. 2009. *The Storm*. London: Atlantic Books.

Cameron, D. 2010. Big Society speech. Available at: http://www.number10.gov. uk/news [accessed: 22 July 2010].

Chu, B. 2011. Prospects bleak apart from Olympics and inflation. *i newspaper*, 26 December.

Coats, D. 2006. No going back to the 1970s? The case for a revival of industrial democracy. *Public Policy Research*, 13(4), 262–71.

Cooper, M.D. 2000. Towards a model of safety culture. *Safety Science*, 36, 111–36.

Department for Food and Rural Affairs. 2005. *Making Space for Water: Taking Forward a New Government Strategy for Flood and Coastal Erosion Risk*. London: Department for Food and Rural Affairs.

Doeringer, P.B. and Piore, M.J. 1971. *Internal Labour Markets and Manpower Analysis*. Lexington, MA: D.C. Heath and Company.

Giles, J. 2008. Patients doing it for themselves. *New Scientist*, 25 October, 34–5.

Hadden, S.G. 1989. *A Citizen's Right to Know: Risk Communication and Public Policy*. Boulder, CO: Westview Press.

Hall, S. and Jacques, M. (eds) 1989. *New Times – The Changing Face of Politics in the 1990s*. London: Lawrence and Wishart.

Harvey, F., Vidal, J. and Edwards, R. 2011. UK nuclear safety review finds 38 cases for improvement. *The Guardian*, 11 October.

Hirst, P. and Zeitlin, J. 1991. Flexible specialization versus post-Fordism: Theory, evidence and policy implications. *Economy and Society*, 20(1), 1–56.

Hutton, W. 1995. *The State We're In: Why Britain Is in Crisis and How to Overcome It*. London: Vintage.

Irwin, A. 1995. *Citizen Science*. London: Routledge.

Irwin, A. 2001. Constructing the scientific citizen: Science and democracy in the biosciences. *Public Understanding of Science*, 10, 1–18.

Irwin, A. and Wynne, B. (eds) 1996. *Misunderstanding Science?* Cambridge: Cambridge University Press.

Krause, S.S. 1996. *Aircraft Safety: Accident Investigations, Analyses and Applications*. New York, NY: McGraw-Hill.

Kunda, G., Barley, S.R. and Evans, J. 2002. Why do contractors contract? The experience of highly skilled technical professionals in a contingent labour market. *Industrial and Labour Relations Review*, 55(2), 234–61.

Marshall, M.G. 1999. Flexible specialization, supply-side institutionalism and the nature of work systems. *Review of Social Economy*. Available at: http://www2.cddc.vt.edu/digitalfordism/fordism_materials/marshall.htm [accessed: 29 March 2006].

Massachusetts Institute of Technology. 2011. *Airline Industry Overview*. Available at: http://mit.edu/airlines/ [accessed: 6 March 2011].

Mooney, C. 2008. Paradigm Sheep. Available at: http://scienceprogress.org/ [accessed: 2 April 2011].

Mythen, G. and Walklate, S. 2008. Terrorism – Risk and international security: The perils of asking 'What if?' *Security Dialogue*, 30(2–3), 221–42.

Nader, R. 1965. *Unsafe At Any Speed*. New York, NY: Grossman.

Perrow, C. 1984. *Normal Accidents: Living With High-Risk Technologies*. New York, NY: Basic Books.

Piore, M.J. and Sabel, C.F. 1984. *The Second Industrial Divide: Possibilities for Prosperity*. New York, NY: Basic Books.

Prince, R. 2010. David Cameron launches his Big Society. *Daily Telegraph*, 18 July.

Rabeharisoa, V. and Callon, M. 1999. *Le Pouvoir des Malades: l'Association Francaise contre les myopathies et la recherché*. Paris: Les Presse de l'Ecole de Mines.

Roxburgh, A. 2010. *A Summary of Ulrich Beck – Risk Society: Towards a New Modernity*. Available at: http://tcs.ntu.ac.uk/books/titles/rs.html [accessed: 20 January 2011].

Sabel, C.F. 1989. Flexible specialization and the re-emergence of regional economies, in *Reversing Industrial Decline? Industrial Structure and Policy in Britain and Her Competitors*, edited by P. Hirst, and J. Zeitlin. Oxford: Berg, 17–70.

Solnit, R. 2007. Detroit arcadia: Exploring the post-American landscape. *Harpers*, July. Available at: http://www.harpers.org [accessed: 25 August 2010].

Smith, V. 2001. *Crossing the Great Divide: Worker Risk and Opportunity in the New Economy*. Ithaca, NY: Cornell University Press/ILR Press.

Stohl, M. 2008. Old myths, new fantasies and the enduring realities of terrorism. *Critical Studies on Terrorism*, 1(1), 5–16.

Weick, K.E. and Sutcliffe, K.M. 2001. *Managing the Unexpected: Assuring High Performance in an Age of Complexity*. San Francisco, CA: Jossey-Bass.

Whitfield, K. 2000. High-performance workplaces, training, and the distribution of skills. *Industrial Relations*, 39(1), 1–25.

Wynne, B. 1996. May the sheep safely graze? A reflexive view of the expert–lay knowledge divide, in *Risk, Environment and Modernity: Towards a New Ecology*, edited by S. Lash, B. Szerszynski and B. Wynne. London: Sage, 501–23.

Zimmerman, A.D. 1995. Toward a more democratic ethic of technological governance. *Science, Technology and Human Values*, 20(1), 86–107.

1

Empowering Emergency Responders

Roger Miles

Introduction

The principles of Integrated Emergency Management (IEM) that codify best practice for the UK's national emergency preparedness have evolved since their adoption in the 1980s (Walker and Broderick 2006). They were recently embraced by the statutory provisions of the Civil Contingencies Act 2004 and incorporated into the supporting guidance that is promulgated by the government. Central to IEM is a command structure with three levels of control, strategic, tactical and operational, which are termed 'Gold, Silver and Bronze'. This hierarchy is used to coordinate the work of the various parties involved in the response to a major emergency. Each tier's responsibilities are specified and stress is placed on the importance of establishing good communications between them. The legislation requires those links to be forged and maintained through meetings and training.

The response to an incident begins with local emergency services and other agencies; IEM arrangements are implemented locally at first, with regional and national effort being introduced as the need for more substantial, concerted effort escalates. Central to this planning is the concept that the response will be delivered primarily by people performing their normal functions in their area of competence but under abnormal circumstances. A senior officer from the area police force will usually assume the role of Gold Commander and chair of meetings of the Strategic Co-ordination Group; each emergency service will designate functional managers as Silver and Bronze leaders of tactical and operational parts of the response. Other agencies appoint representatives to each level, as appropriate.

Since major incidents are fortunately rare, many training exercises are mounted to bring together those who might have to work in harmony during an emergency. After each exercise and certainly after any real incident, reviews are conducted to learn the lessons of that experience and to consider how they could inform improved preparedness. Much has been written about the design of effective training; see for example Borodzicz (2005) on the use of simulation and gaming to prepare for crisis management. Even more study has been devoted to the causes of major incidents and ways of preventing or mitigating such disasters (Breakwell 2007, Wisner *et al.* 2004, Perrow 1999). One particular stream of research has examined the lessons to be learnt from analysing the decisions of commanders who occupied 'the hot seat' at major incidents (Flin and Arbuthnot 2002).

Beck (1992) highlights the complexities of modern society and the inherent risks; his analysis invites us to question the efficacy of crisis response planning and training that could be 'illusions of expertise' (Perrow 1999) and unreal, bureaucratic exercises. Beck's writings call for a radical openness in addressing complex risks that may require a break from past thinking and preparedness for novel action. Thus training designers can be challenged to reflect on what might be lost by concentrating on the command structure and the administrative protocols, perhaps at the expense of considering the diversity of stakeholders' needs. This chapter considers one element that could be inhibited by rigid organisation, namely the promotion of individual enterprise of responders working within the system. To what extent should their abilities and confidence to act flexibly be encouraged?

Learning from Experience

Official inquiries into major incidents usually produce recommendations about legal, technical and managerial systems improvements that should reduce the risks of a recurrence. Important as these are, one must also note that these reports often include something similar to this finding from the second report on the Buncefield oil storage depot fire in December 2005:

> *The impressive response to Buncefield ... relied on initiative and good working relations of the responders in dealing with an incident that had been unforeseen and therefore not planned for. (Buncefield Major Incident Investigation Board 2007)*

Whilst reviewing the response to a major rail accident, the author found that the recovery work was helped significantly by the rapid arrival of certain equipment. This had been readied and dispatched, without higher authority, by a junior manager who said: 'I knew from experience that it would be needed'. Improvisation is often very effective in the post-disaster recovery phase when police and fire officers use local knowledge to get vehicles from farmers, select access routes and mobilise informal community help. Unfortunately anecdotal accounts of such positive contributions are not prominent in the crisis management literature, but experienced professionals will often refer to them.

Many training exercises are designed to develop procedures that should contribute to effective Command, Control and Communications (C3) during an emergency; thus contact lists and resources are tested along with other such procedural matters. However, if the focus is too closely on organisational issues there may be little opportunity for participants to grapple with the uncertainties and complicating factors of a realistic scenario. There needs to be scope for them to experience the ambiguities and doubts that will inevitably confront many of those involved in an actual crisis as it unfolds. Learning lessons from past events is necessary but not sufficient for the development of a robust response strategy. Such preparations also require encouragement of a readiness to recognise the novel features of new, unprecedented problems whose resolution will demand radical, innovative thinking.

One determinant of the likelihood that officers down the command chain will be prepared to act on their own initiative will be the organisational culture created by their superiors. Research into Gold command level disaster response leadership, reported by Devitt and Borodzicz (2008), asserts that such leaders need competence in four areas: task skills, interpersonal skills, awareness of the diversity of stakeholder needs and personal qualities such as confidence and pragmatism. They argue that these four facets of leadership need to be interwoven like the strands of a rope and they suggest that cultural factors could inhibit the flexibility needed to achieve an effective balance.

Experienced emergency services officers have often commented to the author that a strength of the UK's application of Gold, Silver and Bronze (GSB) is that it facilitates autonomous decisions on the ground by those directly engaged with the public. Support to local actions comes from supervisors at the Silver, tactical level and from chief officers who stay away at Gold command, where they have oversight of strategic operations planning and resources. That is how it should be.

Authoritarian Leadership

Whilst GSB should work openly and flexibly, we can anticipate that an authoritarian leader with a narrow focus could interpret the command structure of IEM as a rigid parallel of a military hierarchy, where authorisation must be obtained before any action and orders must be obeyed without question. The inappropriateness of such a style for problem solving within the multi-dimensional complexity of Beck's Risk Society can be seen in the researches reviewed by Devitt and Borodzicz (2008). They found that some '... current models of crisis leadership fail to establish a balance between the requirement for task skills, interpersonal skills, stakeholder awareness and personal qualities of commanders and their teams'. Examination of this contemporary concern can be underpinned by looking back to earlier research into the personality differences between good and bad generals (Dixon 1994).

Working in the 1970s, Dixon analysed historic military operations to identify the psychological factors underlying the characters of the Generals who presided over various successes and failures. His book includes a theme that is very pertinent to this chapter's focus on the importance of effective action at lower levels:

> Over the years military incompetence has resulted more from a dearth of boldness than from a lack of caution, and more from a pall of indecision than from an excess of impulsivity. [Dixon provides compelling examples of:] ... the not infrequent contrast between the verve and initiative of junior commanders and the cautious indecision of those at higher levels of control. (Dixon 1994)

Dixon argued that authoritarian personalities, egocentric and so pre-occupied with their own self-esteem, could prosper in the reassuringly ritualised structure of the armed services by obedient conformity. However, when placed in operational command their self-doubts and rigidity of thinking led to disastrous decision-making, and particularly a resistance to using new technology or adopting innovative ideas.

When first published in 1976, Dixon's book attracted some angry reactions from some military people who did not know that before becoming a Professor of Psychology, Norman Dixon had served for ten years as a Royal Engineers officer and was wounded, as he readily acknowledged, 'largely through my own incompetence'. It reflects well on the British Army that the provocative value of Dixon's research was soon recognised by regular invitations to contribute to the training of young officers. His book was reprinted in 1994.

Everyone involved in preparing disaster response organisations could perhaps benefit from Dixon's (1994) study of the failures of British Generals in the Crimean and South African campaigns that led him to observe that Victorian Great Britain had '… sent out highly regimented armies which endeavoured to make up in courage, discipline and visual splendour what they lacked in relevant training, technology and adequate leadership'.

Writing recently about safety and efficiency in aviation maintenance engineering, Bennett (2010) argues for 'methodical scepticism' that questions established doctrine and is receptive to new ideas that can lead to beneficial change. Bennett uses the example of Major Ord Wingate who in World War II introduced new ways of fighting the Japanese forces deep inside their jungle territory in Burma. Wingate's novel tactics and leadership skills proved very successful; he earned the respect of his opponents. Those in the British Army hierarchy who resented his being odd or different did not appreciate Wingate, but General Slim who took command in 1943 very clearly recognised his contribution. Dixon (1994) writes very positively about Slim's personality and leadership ability; he cites Slim's praise for Wingate as '… another index of Slim's inner strength: his absence of petty jealousy'.

High Reliability Organisations

Another insightful study that is well worth revisiting, is a paper by two political scientists and a psychologist (Rochlin, La Porte and Roberts 1987) describing their observations on US aircraft carriers. Although the ship's command structure was the US Navy's rigid rank hierarchy, Rochlin *et al.* were surprised to see how the highly technical and hazardous flight deck operations were conducted in adaptable, flexible ways without constant passing of messages up and down the chain of command. Specifically they were interested to find that for safety and operational effectiveness reasons:

> *Even the lowest rating on the deck has not only the authority, but the obligation to suspend flight operations immediately, under the proper circumstances and without first clearing it with superiors. Although his judgement may later be reviewed, he will not be penalized for being*

wrong and will often be publicly congratulated if he is right. (Rochlin et al. 1987)

These responsibilities are in the hands of young, junior personnel and their work teams are subject to frequent changes. Thus before each period at sea there must be an intensive 'work-up' training programme for all the crew to come together as an effective unit. Thereafter drills and exercises are part of daily routines that build confidence and flexibility to cope if something goes wrong. Also 'belt and braces' safeguards or redundancies within the technical and management systems provide the duplication and overlaps whereby high levels of operational reliability and robustness under stress are achieved. These measures are the springboard for resilience.

Other studies in stressful, demanding workplaces such as air traffic control centres have underlined the structural importance of flexibility and redundancy in such 'high reliability organisations (HROs)'; a body of academic research and theory on HROs has been developed (Reason 1997). A central figure is Karl Weick who in 1987 asserted that structural organisation must come first – commonly understood procedures and a coherent organisational culture must be established before individuals within the HRO can be enabled to act independently and appropriately in a crisis. He went on to set out the importance of 'collective mindfulness'[1] in the development of the safety culture in an organisation and to make this concept the foundation of a team's professional capacity to anticipate and contain problems (Weick and Sutcliffe 2007).

Sheffi (2005) drew on HRO research to analyse how commercial businesses can reduce the vulnerability of their supply chains by developing a culture of corporate flexibility, with distributed decision-making and good communications at all levels. He offers business leaders many vivid examples of success or failure to deal resiliently with the volatility of today's global marketplace. Sheffi's analysis draws out some strategic management lessons but particularly emphasises the importance of communicating with staff to 'keep everybody on the same page' and of empowering them to initiate flexible, problem-solving actions at a local level. Investment in training is also vital so that teams can '… morph quickly as the rules of the game change …' and so can be ready for big market demand fluctuations or high-impact/low-probability disruptions.

A consequence of our modern, volatile world is that in most organisations staff changes will keep altering the composition of work teams. This will

---

1   To be mindful is to be aware, critical, reflexive and 'in the moment'.

be particularly true of temporarily constituted civilian emergency services teams that have to come together when a rare disaster strikes. Inevitably some of the people and teams will not have worked together before. Hence the wider relevance of military HRO examples that conclusively establish the value of extensive training to counter the effects of inevitable staffing turbulence. They prompt questions as to the frequency and scale of the civil contingencies training that should be given locally for public confidence that an infrequent multi-agency crisis response within the IEM framework would be effective.

Integrated Emergency Management

The UK's adoption and development of IEM was informed early on by an initiative of the US government's Federal Emergency Management Agency in 1983; this brought together academics interested in developing a coherent study of disaster response management principles and practices; a substantial publication resulted from their deliberations (Comfort1988). This collected volume of chapters by experts examined national government policy imperatives and the many organisational requirements needed to establish effective disaster management. It linked studies of risk reduction, damage mitigation and recovery, based on case studies of responses to incidents, with analyses of the processes of organisational learning and development involved in systems improvement. A major outcome was the recognition that pre-planning and organisation is essential as much can be done to strengthen national readiness across the range of services that may be needed. However, it was also clear that the unexpected will happen and so those plans need to retain scope for flexibility by combining rational analysis with creativity.

One of Comfort's chapters (Lewis 1988) argued that IEM should be an organic, not mechanistic, management structure and that:

> ... there will still be situations – particularly in the post impact emergency period – in which lower level personnel are faced with on-site decisions ... for which there are no clear pre-existing policies or standard operating procedures. In fact, some situations may involve 'bending the rules' and some may even require actions that violate existing policies ... an internal climate is needed that encourages individuals to make

> *independent decisions where necessary and enhances their ability to make them in a creative manner. (Lewis 1988:175)*

Thus there is a basis in research and theory for being concerned that whilst training exercises are vital for establishing local area preparedness by developing effective C3, they should also be aiming to recognise and develop the capabilities of individuals to use their experience and skills fully. Going forward from Beck's analysis, it can be argued that particularly for those at the tactical and operational levels of response management, training should foster their confidence to act decisively, and to an appropriate degree independently, because of the local priorities and complications such as breaks in communications that will inevitably arise during many disasters.

The UK government's policy and procedures for promoting IEM were drawn together in a framework document entitled *Dealing with Disaster*; it was first published by the Home Office in 1992 but subsequent editions in 1994 and 1997 incorporated lessons learned from recent incidents and training exercises. The status of this document as a repository of best practice was confirmed when, in support of the major new legislation that was the Civil Contingencies Act 2004, the Cabinet Office reissued the third edition in 2003.

Fundamental to the guidance given in *Dealing with Disaster* is the principle that '… prime responsibility for handling disasters should remain at the local level where the resources and expertise are found' (Home Office 1997). Furthermore, that guidance recognised that many incidents with the potential to escalate into a crisis can be dealt with completely by an effective local response. Thus the competences of the first responders and their supervisors will be crucial to early resolution of many incidents and will be the foundation for how well a major emergency is handled. The whole edifice of the procedural arrangements rests on the abilities of the practitioners on the ground.

Grint (2010) tells us how a useful leadership concept comes from an ancient Chinese story that likened the leadership abilities of the Emperor Liu Bang to a well-built wheel. The spokes represent the collective resources of the community but getting the right spaces between the spokes is critical to the strength and performance of the wheel. Thus the story tells that the Emperor's success in unifying his great country was based on his capacity to provide the spaces that '… represent the autonomy for followers to grow into leaders themselves'.

Recent research provides three constructive ways of thinking about how to promote such individual effectiveness: the 'flow' concept, from the positive psychology movement; the concept of 'mental toughness' from sports psychology; and the 'portal experience' concept, from research with fire-fighters. Each concept provides a way of talking about an important aspect of the challenges faced by emergency response personnel.

The Flow Concept

In 1998 the American Psychological Society's President called for a move away from the established focus on human shortcomings towards a 'positive psychology' that seeks to nurture talent and improve normal life (Seligman 1999). This stimulus has prompted a good deal of research into 'optimal experience, transcendent performance, excellence and positive deviance' (Fullagar and Kelloway 2009). This work makes use of the concept of 'flow', coined by Csikszentmihalyi as long ago as 1975 and defined as follows:

> *Flow has been defined as the experience of working at full capacity with intense engagement and effortless action, where personal skills match required challenges (Nakamura and Csikszentmihalyi 2002). It is regarded as an 'optimal experience' to such an extent that the two terms are often used interchangeably. (Fullagar and Kelloway 2009)*

A challenging activity brings about the 'flow' state because it requires skilled performance, a merging of action and awareness, clear goals and feedback, loss of self-consciousness and concentration on the task in hand with perhaps some transformation of time that includes apparent loss of time awareness (Csikszentmihalyi 1990). All these components can be seen in the work of emergency service personnel during a major incident, where time is short and yet can appear to stand still when information is lacking, conflicting or uncertain.

Media reports from Haiti of the aftermath of the January 2010 earthquake included close-ups of specialist rescue teams burrowing into wrecked buildings (British Broadcasting Corporation 2010). These skilled workers were justifiably elated when they extracted survivors, especially a man found alive 11 days after the earthquake struck. Their pleasure in being able to apply their training so successfully can be seen as vivid examples of the 'flow' experience. We can also recognise that such experience is sometimes referred to as prompting the 'buzz' that emergency services officers can get from applying their training to real incidents. What the flow concept adds to our understanding of those emotions is the importance of feelings of professional efficacy and confidence that can energise personal and team competence.

Mental Toughness

Positive psychology has had a particular impact on the field of sports psychology and the study of the much talked about 'mental toughness' that is needed to succeed in highly competitive sports (Sheard 2010). Mental toughness is conceptually related to an athlete's resilience, including in recovering from injury; it can be measured, developed and maintained through training (Loehr 1995). The components of mental toughness include the 'hardiness' which is often attributed to stress-resistant individuals. The relevance of this work to the development of emergency response personnel can be seen in this reference to one of the major contributors to this topic:

> According to Jim Loehr (1986), mentally tough performers are disciplined thinkers who respond to pressure in ways which enable them to remain feeling relaxed, calm, and energized because they have the ability to increase their flow of positive energy in crisis and adversity. ... Under competitive pressure, mentally tough performers can continue to think productively, positively, and realistically and do so with composed clarity. (Sheard 2010)

A key point from the mental toughness literature is that it is not about promoting reckless bravery or blind obstinacy. Sheard (2010) summarises the support given to a group climbing Everest in 2007 by a psychologist, David Fletcher, who was concerned to temper their high motivation with self-awareness, '... and knowledge of when to re-group and return another day'. This knowledge component is important as it underlines that mental toughness is an attribute that can be learned. Also this example shows that the achievement mindset of the mentally tough includes an important element of judgement as to when securing success may need to be deferred.

In closing this section we should note that during a disaster many of those responding are at risk of high-pressure media scrutiny. They could be exposed to criticism for choices made in much the same way that sporting officials are often challenged to justify decisions or to acknowledge mistakes. Sheard (2010) reports a research finding that 'professional rugby league referees were as mentally tough as the players they were officiating'.

Portal Experiences

Holgate and Clancy (2009) examined the impact of having experienced dangerous situations on the risk perception and attitudes to safety of volunteer firefighters in Australia. They drew on the concept of 'portal experience' developed recently in studies of US firefighters who had experienced actual incidents or near-misses. They found that more than half of their sample of 110 firefighters had had potentially life-threatening experiences, many in the extremely dangerous wildfires that have ravaged areas of Australia in recent years. As a result these experienced firefighters were more alert to the risks in scenarios and were disposed to train more conscientiously. In particular Holgate and Clancy (2009) found that 43 per cent of their 60 firefighters that had had a portal experience said that one change in their approach to fire incidents was: 'I "stand back" more on scene'.

These behavioural changes in approach as a result of portal experiences are attributed to an emotional 'affect heuristic' as conceptualised by Slovic and Peters (2006), and are not thought to be the result of rational thought and analysis. There has been a considerable amount of other research similarly focused on how emotions play an important part in risk perception and risk decision-making in crisis situations (Breakwell 2007).

Specifically relevant to this chapter is Holgate and Clancy's recommendation that improvements in incident reporting systems should be made to capture more personal insights, 'in order to gain training advantage from fire fighters' portal experiences'. This accords with the importance that has long been attached to first-hand case studies in the training of emergency services personnel, and particularly in the multi-agency training provided by the UK government's Emergency Planning College at Easingwold in Yorkshire. However, the emotional component spotlighted in this research underlines that such training should not be a dry, intellectual knowledge transfer but should be offered in ways that get across the feelings and attitude changes experienced.

Discussion

One way of bringing together these three concepts from research is to view skilled emergency responders as 'craftsmen'. We can then utilise that term's richness as presented by Sennett (2009) in his philosophical review of skilled work, past and present, and the standing of craftsmanship in modern society. Sennett shows us that craftsmen are all of the following:

- skilled performers, able to visualise and mentally rehearse options;

- highly motivated;

- confident of their self-efficacy or 'can do' abilities;

- time transformers, able to control their time awareness and so avoid spiralling into panic under time pressure;

- resilient, and able to reframe problems and bounce back.

Sennett (2009) argues that personal motivation to do good work is inseparable from social organisation and can be enhanced by a sense of vocation, but that pride and competition for recognition can work negatively. Thus organisations need to be careful about their strategies for harnessing that motivation so that their master craftsmen are in his words, 'sociable experts'. Sennett also asserts that building craftsmanship and strengthening individual skills will be a very useful strategy for dealing with the dislocations that are inevitable consequences of the complexities of modern society.

The craftsman's capacities to overcome snags and anticipate new possibilities fits well with the basic premise of IEM that an effective emergency response will be based on police officers and others doing their normal jobs in exceptional circumstances: 'Emergency plans must build on routine arrangements' (Home Office 1997). Or as Perrow (1999) points out, functional plans are made possible by affinities that are extensions from the known, but as discussed earlier, they need flexibility and openness to tackle the unforeseen. Thus UK government planning has a generic basic structure so that the GSB command and control system can be applied to a wide range of circumstances and can automatically engage different components of the particular response, according to the unique needs of the actual crisis situation (Home Office 1997).

Delegation of responsibilities is key to that bespoke tailoring being achieved. For example, it will be local response personnel who will know who are the official and more importantly the unofficial leaders within the various social groups in the vicinity. The handling of fatalities at the scene can present major problems if different faith communities are involved; contacting leaders of those groups will be best done by local officers who are known and trusted because of established relationships.

Such devolved authority will be attended by obligations; in particular there will always be a duty to report regularly to higher command. Also local decisions will need to be recorded so that subsequent inquiries can follow a clear audit trail. Thus opening logs and ensuring that contemporary record-keeping is maintained are widespread responsibilities, including at the operational level. Everyone who takes an initiative to deal with a local issue must expect some form of post-event scrutiny, perhaps in a coroners court or police investigations leading to criminal court proceedings.

Another significant lesson from many major emergencies is that the affected population will respond to events autonomously and that adaptive behaviour of citizens should therefore be anticipated in emergency response planning. Wisner *et al.* (2004) provide a potent discussion of this reality, which includes lessons learned from the Kobe earthquake in 1995 where the community response was markedly faster than that of the national government. This general point of community resilience was taken up by Boin and McConnell (2007) in their arguing for disaster response planning to be localised rather than top down, and that front-line responders should be ready to tailor their actions accordingly. Grint (2010) also discusses how local communities might be encouraged to develop 'responsible citizenship that is more likely to engage in acts of leadership', but recognises that a balance is needed between long-term institutionalised leadership and temporary local decision-making.

Thus the organisational design for application of IEM policy within a community area should provide a framework for action that recognises and builds on the capabilities of available staff. Within the overall strategy procedural details should emerge from assessment of those capabilities and experience of effective team performance in training and practice. Clegg and Spencer (2007) argued for such an emergent rather than 'top down' approach to the design of work when they proposed a circular relationship between job characteristics and motivational states. They drew on researched examples to show that feedback affects the nature of the job and to contrast this with the traditional view that job characteristics act linearly from job to motivational state to job outcome. Furthermore, they asserted that when people can alter or craft their roles to suit their own personal needs then they can create a more satisfying job and so be motivated to perform better.

Conclusion

National Government's strategic emergency planning policy sets the framework and provides the resources for society to respond to a disaster. However, the quality of the operational response will also depend on the actions of operational teams and the skilled leadership provided by team leaders and supervisors at each level of the prescribed Command and Control system. Thus, human abilities and effective social interactions will be the lifeblood of success, not the pre-determined organisational protocols and procedures laid down in the operational frameworks and policies. In order to foster those abilities the importance of the craftsmanship of those individuals needs to be recognised in the planning of procedures and also nurtured in frequent training exercises. Military examples were used in this discourse to spotlight the extent and frequency of training that should be provided for effective work in crisis situations. Concepts from recent researches in different fields of psychology and sociology have been used to similarly identify qualitative features that such training should offer. It has been argued that these are needed in order to maximise the capabilities of individual emergency responders to act confidently and flexibly when confronted by the pressing, novel problems that a new disaster will inevitably present to them.

References

British Broadcasting Corporation 2010. Reports from Port au Prince. *BBC 24 hour News*, 23 January.

Beck, U. 1992. *Risk Society: Towards a New Modernity*. London: Sage.

Bennett, S.A. 2010. Human Factors for Maintenance Engineers and Others – A Prerequisite for Success, in *Encyclopaedia of Aerospace Engineering*, edited by R. Blockley and W. Shyy. Chichester: John Wiley & Sons.

Breakwell, G.M. 2007. *The Psychology of Risk*. Cambridge: Cambridge University Press.

Boin, A. and McConnell, A. 2007. Preparing for Critical Infrastructure Breakdowns: The Limits of Crisis Management and the Need for Resilience. *Journal of Contingencies and Crisis Management*, 15(1), 50–59.

Borodzicz, E.P. 2005. *Risk, Crisis and Security Management*. Chichester: John Wiley & Sons.

Buncefield Major Incident Investigation Board 2007. *Recommendations on the Emergency Preparedness for, Response to and Recovery from Incidents*, July.

Clegg, C. and Spencer, C. 2007. A Circular and Dynamic Model of the Process of Job Design. *Journal of Occupational and Organizational Psychology*, 80, 321–39.

Comfort, L.K. (ed) 1988. Managing Disaster: Strategies and Policy Perspectives. London and Durham, NC: Duke University Press.

Csikszentmihalyi, M. 1990. *Flow: The Psychology of Optimal Experience*. New York: Harper.

Devitt, K.R. and Borodzicz, P. 2008. Interwoven Leadership: The Missing Link in Multi-Agency Major Incident Response. *Journal of Contingencies and Crisis Management*, 16(4), 208–16.

Dixon, N. 1994. *On the Psychology of Military Incompetence*. London: Pimlico.

Flin, R. and Arbuthnot, K. (eds) 2002. *Incident Command: Tales from the Hot Seat*. Aldershot: Ashgate.

Fullagar, C.J. and Kelloway, E.K. 2009. 'Flow' at Work: An Experience Sampling Approach. *Journal of Occupational and Organizational Psychology*, 82, 595–615.

Grint, K. 2010. *Leadership: A Very Short Introduction*. Oxford: Oxford University Press.

Holgate, A. and Clancy, D. 2009. Portal Experiences: The Impact of Firefighters' Experiences of Threat on Risk Perception and Attitudes to Personal Safety. *Australian Journal of Emergency Management*, 24(3), 15–20.

Home Office 1997. *Dealing with Disaster*. Liverpool: Brodie Publishing.

Lewis, R.G. 1988. Management Issues in Emergency Response, in *Managing Disaster: Strategies and Policy Perspectives*, edited by L.K. Comfort. London and Durham, NC: Duke University Press.

Loehr, J.E. 1995. *The New Toughness Training for Sports*. New York, NY: Plume.

Nakamura, J. and Csikszentmihalyi, M. 2002. The Concept of Flow, in *Handbook of Positive Psychology*, edited by C.R. Snyder and J.S. Lopez. New York: Oxford University Press.

Perrow, C. 1999. *Normal Accidents: Living with High-Risk Technologies*. Princeton, NJ: Princeton University Press.

Reason, J. 1997. *Managing the Risks of Organisational Accidents*. Aldershot: Ashgate.

Rochlin, G.I., La Porte, T.R. and Roberts, K.H. 1987. The Self-Designing High-Reliability Organisation: Aircraft Carrier Flight Operations at Sea. *Naval War College Review*, Autumn, 76–90.

Seligman, M.E.P. 1999. The President's Address. *American Psychologist*, 54, 559–62.

Sennett, R. 2009. *The Craftsman*. London: Penguin Books.

Sheard, M. 2010. *Mental Toughness: The Mindset Behind Sporting Achievement*. Hove: Routledge.

Sheffi, Y. 2005. *The Resilient Enterprise: Overcoming Vulnerability for Competitive Advantage*. Cambridge, MA: MIT Press.

Slovic, P. and Peters, E. 2006. Risk Perception and Affect. *Current Directions in Psychological Science*, 15(6), 322–25.

Walker, C. and Broderick, J. 2006. *The Civil Contingencies Act 2004: Risk, Resilience and the Law in the United Kingdom*. Oxford: Oxford University Press.

Weick, K.E. 1987. Organisational Culture as a Source of High Reliability. *California Management Review*, 29, 112–27.

Weick, K.E. and Sutcliffe, K.M. 2007. *Managing the Unexpected: Resilient Performance in an Age of Uncertainty (Second Edition)*. San Francisco, CA: Jossey-Bass.

Wisner, B., Blaikie, P., Cannon, T. and Davis, I. 2004. *At Risk: Natural Hazards, People's Vulnerability, and Disasters (Second Edition)*. London: Routledge.

2

Terrorism and the Risk Society

David Waddington and Kerry McSeveny

Introduction

> *From the vantage point of the Brooklyn Heights, we saw Lower Manhattan disappear into dust. New York, and therefore all cities, looked fragile and vulnerable. The technology that was bringing us these scenes has wired us closely together into a febrile, mutual dependency. Our way of life, centralised and machine-dependent, has made us frail. Our civilisation, it suddenly seemed, our way of life, is easy to wreck when there are sufficient resources and cruel intent. No missile defence system can protect us.*
>
> *Yesterday afternoon, for a dreamlike, immeasurable period, the appearance was of total war, and of the world's mightiest empire in ruins ... Like millions, perhaps billions around the world, we knew we were living through a time that we would never be able to forget. We also knew, though it was too soon to wonder how or why, that the world would never be the same. We knew only that it would be worse. (McEwan 2001)*

The above account by the award-winning novelist, Ian McEwan, recalls in a graphic and chilling manner the way in which many of us looked on in horror at the televised spectacle of a pair of hijacked planes careering with massively destructive impact into the New York World Trade Centre on 11 September 2001. Occurring in conjunction with a simultaneous suicide attack in which two further hijacked aircraft were directed at the Pentagon, the '9/11' atrocity was disconcertingly without precedent:

> *The enormity and sheer scale of the simultaneous suicide attacks on September 11 eclipsed anything previously seen in terrorism. Among*

> *the most significant characteristics of the operation were its ambitious*
> *scope and dimensions; impressive coordination and synchronization;*
> *and the unswerving dedication and determination of the 19 aircraft*
> *hijackers who willingly and wantonly killed themselves, the passengers,*
> *and crews of the four aircraft they commandeered and the approximately*
> *3,000 persons working at or visiting both the World Trade Center and*
> *the Pentagon. (Hoffman 2002)*

The shocking impact of 9/11, and a series of subsequent atrocities – such as the '7/7' (7 July 2005) London suicide attacks in which 56 people, including four bombers, were killed in three explosions on the underground and one aboard a double-decker bus – have consolidated the development of a *New Terrorism* discourse in academic and political circles, which depicts contemporary terrorism as fundamentally different from more traditional forms, epitomized by the late-twentieth century activities of (say) ETA and the IRA (Mythen and Walklate 2006a).

Basic to this discourse is the notion that modern terrorism is no longer concentrated on any given locality, or underpinned by a particularly coherent and unifying political rationale; rather, 'new terrorist groups are defined by their amorphous aims, disparate organization and capacity to strike across different continents' (Mythen and Walklate 2006a). These, and related suppositions, that the 'new' terrorists (typified by 'extreme Muslim fundamentalist' groups like al Qaeda) are 'inspired by religious extremism and ethnic separatism', and have both the will and potential to deploy weapons of mass destruction, have engendered a 'new' security environment in which international terrorism 'has replaced the Cold War as the principal conflict threatening the integrity of Western liberal states' (Pantazis and Pemberton 2009).

The post-9/11 era was one in which successive UK New Labour governments under Tony Blair and his successor, Gordon Brown, were outspoken advocates of the New Terrorism discourse. The former told the World Affairs Council in 2006, for example, that global acts of terrorism were being undertaken by 'A movement that believed Muslims had departed from their proper faith, were being taken over by Western culture, were being governed treacherously by Muslims complicit in this take-over, whereas the true way to recover not just the true faith, but Muslim confidence and self esteem, was to take on the West and all its works' (Blair 2006). Mr Blair's views clearly resonated with those of senior UK intelligence personnel. Speaking shortly before her retirement, a former director of the British Security Service observed that the threat now posed by al Qaeda and similar terrorist organizations was both global in reach and 'unprecedented in scale, ambition and ruthlessness', and that it remained 'a very real possibility that they may, some time, somewhere, attempt a chemical, biological, radiological or even nuclear attack' (Manningham-Buller cited in Croft and Moore 2010).

It will come as no surprize to discover that developments in contemporary international terrorism have been closely considered by Ulrich Beck, with regard to his theory of Risk Society (for example, Beck 1992, 1998, 2002, 2006, 2009). This chapter provides an overview of Beck's approach, together with an acknowledgement of its shortcomings and an outline of a complementary and, in some ways, competing perspective. The chapter begins with a brief summary, followed by a preliminary critique of Beck's understanding of the apparent intensification of world terrorism. A pair of subsequent sections then set out to develop two particular aspects of Beck's work on terrorism – namely: the role of political and media manipulation in framing the public's awareness and understanding of the phenomenon, and the way in which such impression management becomes translated into misguided and counter-productive domestic and foreign security policies. The penultimate section reviews evidence to suggest that UK security measures, both on the home front and abroad, have recently been predicated on a flawed understanding of the apparent motivation for terrorist activity. The conclusion therefore advocates a radical rethink of the British approach to eradicating contemporary terrorism.

Beck's Approach to Terrorism

Other chapters in this volume have emphasized as a core tenet of Beck's Risk Theory his position that, whereas industrial societies had hitherto been prone to the risk of natural disasters, such as earthquakes, famine and flood, the onset of capitalist modernization has been accompanied by a far greater prevalence of 'manufactured risks' – e.g. of harmful climate change, global financial crisis, transnational crime and worldwide terrorist activity.

The recent growth and ubiquity of international terrorism has three major features in common with the other types of manufactured risk that Beck refers to. First, the localized risks associated with earlier industrial society have become 'borderless' or 'de-territorialized', and potentially harmful to all the world's nations. As Beck (1998) himself succinctly explains, 'there are no bystanders anymore'. Secondly, techno-scientific developments have helped to ensure that the above types of risk carry a progressively increasing potential to inflict great and catastrophic harm. Suicide bombs, airplane hijacks and strikes on military installations may be one thing, but, in addition, 'Every advance from gene technology to nanotechnology opens a "Pandora's box" that could be used as a terrorist's toolkit' (Beck 2002). Finally, such risks are so unpredictable, both in timing and location, as to be practically impossible to be effectively protected or insured against (Beck 2002).

Whilst environmental and financial problems associated with 'modernity's' tendency towards self-endangerment typically constitute unintentional side effects of the world manufacture of goods and services, terrorist activity is, in Beck's terms, 'intentionally bad', in that 'it aims to produce the effects that the other crises produce unintentionally' (Beck 2002).

It is, according to Beck (2009), the nebulous structure of the terrorist organization, the unpredictability of its activities and intangible quality of its goals that make it so difficult to thwart or guard against:

> The murky terrorist networks are, as it were, 'violence NGOs'. Like non-governmental organizations in civil society, they operate independently of territory, without an organizational centre, hence at once locally and transnationally ... They suddenly attack where, until then, nobody had expected them to. They demonstrate their power of destruction by transforming civic social spaces into potential death zones. Their terror is indiscriminate, aimless and unpredictable. No battle takes place. The threat is in the most radical sense asymmetrical. For the victims, whether in uniform or not, have no opportunity to resist. Courage is as futile as cowardice. The terrorists' primary weapon is fear. They do not want to gain a victory but to create panic.

The indignant and even alarmist responses by senior world leaders tend, ironically, to have the unintended consequences of not only encouraging the terrorists to believe in their capacity to injure and unsettle even the world's foremost superpower (Beck 2002), but also to guarantee that the perpetrators will suddenly rise to 'terrorist world stardom' (Beck 2009). Moreover there is a risk that, in the very act of anticipating and highlighting the nature of future atrocities, nation states will be transmitting those very ideas and opportunities to their would-be terrorist assailants (Beck 2009).

Beck's writings have shown an increasing awareness of the extent to which perceptions of risk are prone to being 'magnified, dramatized or minimized' by key ideological institutions, notably the mass media, and ultimately the State itself (Beck 1992). 'The main question', as he puts it, 'is who defines the identity of a "transnational terrorist"? Neither judges nor international courts, but powerful governments and states. They empower themselves by defining who is their terrorist enemy, their bin Laden' (Beck 2002). Within this process there is a tendency for the distinction between rationality and hysteria to disappear. Though confessing himself afraid to 'even dare think about deliberate attempts to instrumentalize this situation' (Beck 2006), Beck (2002) was nonetheless sensitive to the possibility that:

> *In order to broaden terrorist enemy images, which, to a large extent, are a one-sided construction of the powerful US state, expanded parameters are being developed so as to include networks and individuals who may be connected to Asian and African terrorist organizations. This way, Washington constructs the threat as immense. Bush insists that permanent mobilization of the American nation is required, that the military budget be vastly increased, that civil liberties be restricted and that critics be chided as unpatriotic.*

Beck recognizes that one especially unsavoury consequence of the enhanced perception of terrorist threat is its negative impacts, both on human rights and the readiness to make 'pre-emptive', retaliatory or retributive strikes against other nations:

> *With danger, what saves also grows us – because, faced with the alternative 'freedom or security', the vast majority of human beings seem to prioritize security, even if that means civil liberties are cut back or even suffocated. As a result of the experience of the risk of terror, there is an increasing readiness, even in the centres of democracy, to break with fundamental values and principles of humanity and modernity, e.g. with the principle 'There can be no torture' or 'Nuclear weapons are not for use', that is, to globalize the practice of torture and to threaten so-called 'terror states' with a preventative nuclear strike. (Beck 2006)*

Host nations are also likely to subject to increased levels of surveillance and legal control those sections of society it regards with enhanced but – invariably unwarranted – feelings of suspicion and mistrust. To use Beck's (2009) own succinct turn of phrase, 'Risk divides, excludes and stigmatizes'.

For Beck, the solution to stemming the almost inevitable tide of world terrorism resides in the paradox that global terrorism has unintentionally created a case for greater transnational cooperation between extended networks of nation states (Beck 2002). He therefore urges the major world powers to adopt a more humanitarian and inclusive 'cosmopolitan perspective' or 'cosmopolitan imagination' (Beck 2002), which makes 'the negotiation of contradictory cultural experiences', central to all aspects of governmental strategy, both at home and abroad (Beck 2002). The likes of Britain and the United States must therefore aspire to become truly 'Cosmopolitan states' and show a readiness to abandon their current 'cyclopean vision' in favour of a less parochial focus on issues of global concern. As Beck (2002) explains:

> Cosmopolitan states struggle not only against terror, but against the causes of terror. They seek to regain and renew the power of politics to shape and persuade, and they do this by seeking the solution of global problems that are even now burning humanity's fingertips but which cannot be solved by individual nations on their own. When we set out to revitalize and transform the state in a cosmopolitan state, we are laying the groundwork for international cooperation on the basis of human rights and global justice.

A Critique of Beck

Beck's application of his 'Risk Society' approach to contemporary international terrorism is open to a number of possible criticisms – especially in relation to his conceptions of the possible goals of terrorist groups, the scope of world terrorist activity, and the actual or perceived severity and destructiveness of such behaviour.

Contra Beck's notion that terrorism is 'indiscriminate, aimless and unpredictable', it remains apparent that such activity has been remarkably consistent in terms of its specific targets and possible underlying goals. For example, Spencer (2006) argues that proponents of the New Terrorism discourse not only overlook the fact that fanatical religious terrorism has been in evidence for literally thousands of years, but also fail to acknowledge that 'the distinction between religious and politically motivated terrorism is predominantly artificial'. This point is reinforced by Stohl (2008), who maintains that the assumed association between suicide terrorism and religious fundamentalism is invariably misread or overstated. Thus, as he points out, 'What nearly all suicide terrorist attacks actually have in common is a specific secular and strategic goal: to compel modern democracies to withdraw military forces that the terrorists consider their homeland'.

Common sense further suggests that Beck's focus on the 'borderless' nature of contemporary terrorist activity obscures the fact that the threat of terrorism continues to focus on a relative *handful* of international locations and does not constitute the same kind of enduring possibility for the unaffected majority. In other words:

> *Although the 'new terrorist' threat may be potentially universal, in practice certain countries – largely those with histories of economic, cultural, political and religious imperialism – are more endangered than others. We may not be completely surprised if the UK or the USA were subjected to future attacks by Islamic fundamentalist terrorist groups, but it may puzzle us if Slovenia were. (Mythen and Walklate 2008)*

It also seems reasonable to recommend that Beck has overstated what Mythen and Walklate (2006a) refer to as the 'extraordinariness' of terrorism, insofar as the devices used by modern terrorists continue to be 'low-tech and relatively crude' (Mythen and Walklate 2008), and have not so far involved the dreaded weapons of mass destruction (Duyvesteyn 2004). It remains true that most instances of terrorism involve few civilian casualties, and that claims of a sudden rise in 'mass-casualty attacks' is predicated on a relatively few, high-profile examples (Stohl 2008). The 'shock factor' inherent in such activities may well be related, as Spencer suggests, 'to the need of keeping the media and the world's awareness focused on their grievances'. Terrorism is, as he points out, 'still theatre, just on a much bigger stage, where an act has to be big and shocking to keep the audience's short attention from drifting to other scenes'.

Finally, Mythen and Walklate (2006a) fear that Beck is overplaying the transformatory potential of modern terrorism to induce greater cooperation between the major powers and those nations harbouring historical feelings of injustice towards them. They assert that, far from encouraging a spirit of dialogue and reconciliation, the United States and British foreign policies have been predicated on the apparent desire for punishment, vengeance and retribution.

There are two further aspects of Beck's work on terrorism that, without being flawed and, therefore, open to criticism, are each only superficially touched upon but are important enough to warrant further exploration. These aspects of his argument relate to:

a)   the manner in which major Western political elites have 'spun' an essentially self-serving and disingenuous explanation of the nature and underlying motives of international terrorism; and

b)   the way that such spin has been used to justify inappropriately coercive and escalatory foreign and domestic security measures.

These two crucial issues which Beck hints at without ever dealing with in adequate depth are now addressed in turn.

The Political Spin on Terrorism

Mythen and Walklate (2008) make the important point that societal discourses relating to terrorist activity do not emerge 'organically' or by accident; rather they are deliberately constructed by the government and other dominant institutions, notably the police and media. Following Furedi (2005) these authors assert that both British and American governments have constantly striven to maintain a 'climate of fear' around terrorism, thereby keeping it firmly on the political agenda, even when there has been no obvious threat of its occurrence. Mythen and Walklate are not alone in believing that the terrorist discourse has been assiduously propagated on both sides of the Atlantic in justification of highly contentious domestic and foreign policies.

One of the more controversial variations on this perspective is Altheide's (2007) argument that the US administration presided over by George W. Bush ceaselessly manipulated media representations of terrorism as part of its long-term goal to ensure American military pre-eminence and world domination, and nullify potential new challengers. Thus, for example:

> *Bringing about a 'regime change' in Iraq was part of a plan for the United States to become a hegemon, including withdrawing from – if*

> not negating – certain treaties (e.g. nuclear test ban) and becoming
> more independent of the United Nations. The US invasion of Iraq
> was justified, in the main, by claims that Saddam Hussein possessed
> 'weapons of mass destruction' (WMD), was in league with the terrorists
> who attacked the US, and that he was likely to place these weapons at
> the disposal of other terrorists. It took less than a year for the world
> to learn that none of these assertions were true, and indeed, there was
> strong evidence that members of the Bush administration were quite
> aware that such WMDs did not exist.

According to Altheide, prominent members of the Bush administration, including his Vice-President, Dick Cheney, were aligned to the so-called Project for the New American Century (PNAC) group of senior Republican politicians which had drawn up a 'blueprint for US world domination' in 1992. Such figures had allegedly been awaiting 'a catastrophic event, a new "Pearl Harbour," that could be used as a catalyst to adopt a more aggressive foreign policy'. The attacks of 9/11 perfectly fulfilled this function.

Accordingly, 'The next 18 months were spent preparing public opinion for the invasion of Iraq on 20 March 2003. This preparation included the freedom to define and use terrorism in a very broad and general way' (Athleide 2007). A generally compliant American media thereafter reported the war with Iraq in such ways as to demonize and dehumanize the enemy, justify torture as a regrettable but necessary evil and lament the loss of American lives as the inevitable price of freedom from tyranny. Any dissenting media or academic opinion was savagely pilloried as 'misguided, cowardly or unpatriotic'.

The government continued to employ this terrorism discourse to justify its interventions in Iraq and, subsequently, Afghanistan. Mythen and Walklate (2008) report how, in February 2006 for example, President Bush disclosed that the American security agencies had succeeded in thwarting an Al-Qaeda plot to crash an airplane into the 73-storey Library Tower in Los Angeles. However, 'It later transpired that this planned attack, allegedly to be executed by members of Jemaah Islamiah, was uncovered by intelligence services back in 2002' (Mythen and Walklate 2008).

This 'war on terrorism' narrative – backed up by President Bush's warning to all world governments that 'Either you are with us or you are with the terrorists' (Bush cited in Croft and Moore 2010) – had obvious implications for British foreign policy and its justificatory political discourse:

> *America had declared the new age: and that declaration would structure security and foreign policy for the world. For the UK, that would mean that counterterrorism policy, broadly expressed, would be framed far more by the American agenda than it would be by Britain's own experiences in struggles with terrorists. (Croft and Moore 2010)*

In aligning its own foreign and domestic policies so closely to those of the United States, the UK government under Prime Minister Tony Blair also repeatedly employed the New Terrorism discourse in justification of military intervention and tighter national security measures. Thus, for example:

> *The process of leaking by government sources has been particularly notable in relation to possible strikes on the British mainland, with the media reporting a series of plots hatched by fundamentalist Islamic networks. These foiled and allegedly include the crashing of a plane into Canary Wharf Tower; the launching of surface-to-air missiles at Heathrow; explosive strikes on the Houses of Parliament and the detonation of a bomb at Old Trafford football stadium. It remains unclear how accurate the intelligence that generated these claims was, nor is it clear how, where or why this information was passed on to journalists by the government or security services. Nevertheless, it is probable that each revelation of a foiled plot – be it based on credible information or otherwise – has potentially served to ratchet up levels of public concern about 'new terrorism'. (Mythen and Walklate 2008)*

Misguided Policy and its Consequences

Mythen and Walklate (2008) make the important point that discourses of terrorism and risk have the pernicious effect of ensuring individual compliance to tighter patterns of self-regulation, encouraging the tendency to be more suspicious of marginalized groups, and making people more accepting of the need to subject them to closer surveillance and control. The knock-on effect of all this has been the production of a novel raft of counter-terrorism measures – e.g. the Terrorism Act 2000, Anti-Terrorism, Crime and Security Act 2001, and the Prevention of Terrorism Act 2005 – which have made 'suspect populations' out of fearful Muslim communities.

One especially controversial use of police power has been their use of sections 44 and 45 of the Terrorism 2000 Act, which has 'afforded [them] unfettered discretion to stop and search in relation to terrorist activities' (Pantazis and Pemberton 2009). While the numbers of Muslim individuals stopped and searched by the police has rocketed, disproportionately to the non-Muslim UK population (Frost 2008), the relevant encounters rarely result in arrests or convictions being brought, prompting academic speculation that the hidden agenda is that of intelligence gathering (Pantazis and Pemberton 2009). This possibility is acknowledged by the Metropolitan Police Service themselves, who stated in the context of evidence submitted to the Parliamentary Affairs Committee that:

> Section 44 powers do not appear to have proved an effective weapon against terrorism and may be used for other purposes. It has increased the level of distrust of police. It has created deeper racial and ethnic tensions against the police. It has trampled on the basic human rights of too many Londoners. It has cut off valuable sources of community information and intelligence. It has exacerbated community divisions and weakened social cohesion. (Metropolitan Police Service cited in Mythen and Walklate 2009)

This stiffening up of relevant legislation is fundamental to the government's CONTEST (counter-terrorism) strategy, the four components of which are Pursue, Prevent, Protect and Prepare. The relatively soft-centred Prevent (Preventing Violent Extremism or PVE) dimension constitutes a 'hearts and minds' approach with the objectives of:

> increasing the resilience and addressing the grievances of communities, and on indentifying vulnerable individuals, as well as challenging and disrupting ideologies sympathetic to violent extremism ... Here, 'resilience' can be understood as resisting the appeal of, or even standing up to, extremist political activity and terrorist recruitment attempts within Muslim communities. (Thomas 2010)

Prevent-related schemes and mechanisms are administered primarily via educational and welfare agencies and through community organizations. As Thomas explains, the strategy is firmly underpinned by a 'values-based' understanding which takes issue with the way that Islam is interpreted and practised by the younger generations of Muslims, and seeks to propagate a 'more moderate and progressive British variation of the faith' (Thomas 2010). One prominent illustration of this approach is the so-called Channel programme, which provides practical and psychological support for young people identified to the police (by their teachers, for example) as being 'at risk of extremism' (Kundnani 2009).

Thomas (2010) further discloses how relevant funding has been deployed with the intention of promoting the development and influence of new organizations subscribing to a 'more modern and moderate' version of Islam. In contrast, financial support has been withheld or withdrawn from those organizations who, like the Muslim Council of Britain (MCB), are regarded as too hard-line or not sufficiently sympathetic to the anti-terrorist strategy of the British government.

Thomas (2010) maintains that:

> *The danger of this 'values-based' approach, and the fact that funding is contingent on its acceptance, is that it closes down the debates and involvements needed to undermine the appeal of violent extremism.*

Providers and potential end users of Prevent have been critical of the fact that it is primarily limited to Muslim communities, implicitly identifying them as the 'terrorist' or 'extremist' threat to be dealt with. Young Muslims correspondingly regard Prevent funding as 'dirty money', while local authority youth workers and school teachers complain of feeling under increasing pressure to provide information on individuals they are working with. Great scepticism also surrounds the fact that many educational and awareness promotion schemes are led by police officers (Kundnani 2009).

There is concrete evidence that the government's inclination to financially support ostensibly moderate Muslim groups, notably the Sufi community, has cost them the co-operation and support of Salafi and other, arguably more 'street-wise' and 'street-credible' groups, the upshot being that the flow of information has dried up and the police capacity to counteract Al Qaeda propaganda correspondingly diminished (Lambert 2008, Pantazis and Pemberton 2009, Spalek and Lambert 2008). Moreover, its avowedly 'monocultural emphasis' has resulted in what Thomas (2009) refers to as 'two-way envy and resentment, with Muslim communities asking why "extremism", including its violent political form of far-right activists, was not being addressed in some white communities, while non-Muslims questioned why such significant public resources were being directed towards often bland and generalized youth and community activities for Muslims only'. Finally, the perceived convergence of government policy and the 'anti-Islam and anti-immigrant' programme advocated by the BNP has meant that sensations of shame once associated with support for the BNP have evaporated and that the party has been able to make previously unheard of electoral gains (Kundnani 2009).

Debunking the 'Radicalization' Thesis

If it is true, as above sections of this chapter suggest, that Beck has not really attempted to explain the underlying motivation for terrorist activity, and that major Western governments, notably the USA and UK, have wilfully propagated a dangerously misleading account of its socio-political origins, we are still left with the task of trying to better understand the process of radicalization which appears to be the driving force of modern terrorism. The evidence reviewed in this penultimate section helps to cast doubt on the 'New Terrorism' argument that 'Islamic Fundamentalism' or 'religious fanaticism' is primarily to blame. Rather, it seems to chime with the views of those critics of the New Terrorism discourse, like Duyvesteyn (2004) and Spencer (2006), who see the occupation and oppression of foreign states as a prime motivating factor.

Sivanandan (2006) strongly believes that the type of misguided security policies reviewed in the above section have created a widespread 'siege mentality' within Muslim communities and reinforced exactly the type of segregation that the government was purporting to discourage. In his view, the Prime Minister, Tony Blair's insistence on regarding multiculturalism as a possible cause of terrorism – in that individuals steeped in their own culture tended to feel more alienated from mainstream British society – stemmed from a misguided refusal to accept that the war in Iraq was a major causal factor in the suicide bombings.

As both Sivanandan (2006) and Kundnani (2007) point out, it is indisputable that the individuals most prominently involved in the 7/7 suicide bombings were all thoroughly well integrated into mainstream British society, most of them having lived in 'mixed' neighbourhoods and enjoyed good educations. One of those concerned was married to a white woman; a second taught children with learning disabilities from a variety of religions; yet another helped out in his father's chip shop; and the leader of the group (Mohammad Sidique Khan) was a highly gregarious primary school teacher whose colleagues and students universally described him as 'Anglicised'. 'And yet', as Sivanandan (2006) further argues:

> (T)hey were prepared to take their lives and the lives of their fellow citizens in the name of Islam. One reason, therefore, must be as Mohammad Sidique Khan stated it: the invasion and destruction of Iraq. Even by a process of elimination, it is clear that whatever the prize for martyrdom in the hereafter, its cause must be sought in the degradation of Muslim life in the here and now – in Afghanistan, Iraq, Palestine, Bosnia, Chechnya.

This viewpoint is consistent with that expressed by a former head of MI5, who told the Iraq Inquiry how the war had:

> radicalised … a whole generation of young people, some British citizens – not a whole generation, a few among a generation – who saw our involvement in Iraq, on top of our involvement in Afghanistan, as being an attack on Islam. (Manningham-Buller 2010)

Empirical support for this position is provided by two recent studies involving interviews and focus group discussions with British youths of Asian heritage.

Mythen and Walklate's (2009) focus group sessions with 32 British Muslims of Pakistani heritage, aged 18 to 26 years, revealed a collective sense of indignation concerning the way in which Muslim society was represented by the British media as 'unruly, threatening and violent', was greatly at odds with their inherently law-abiding nature. Few respondents were able to identify with or support terrorist activity, but the vast majority could nonetheless sympathize with the underlying sense of grievance:

> Whilst indiscriminate terrorist attacks were rejected, the participants in our study were united in their belief that legitimate and unresolved political, religious and cultural grievances still exist and that these grievances are fundamental in catalyzing the violent actions of 'radicalized' Muslims. In particular, the military incursions into and occupations of Iraq and Afghanistan were frequently cited in talk alongside the intransigence of the British state in challenging atrocities enacted by the Israeli state in Palestine. It was felt by many that the invasion of Iraq was a lop-sided response to the terrorist attacks of 9/11 and that these actions were driven by wider political and economic goals for the US and UK governments. (Mythen and Walklate's 2009)

In a slightly more recent study, Güney (2010) conducted interviews and/or focus group sessions with a total of 23 16–20 year-old Muslim males of Asian heritage. It was evident from these discussions that the youths had developed 'feelings of attachment to an imagined global Muslim community', which had been encouraged and reinforced by the 'perceived' partiality of British media representations of the conflicts in Afghanistan, Iraq and Palestine, and (especially) the loss of Palestinian lives resulting from Israel's military attack on Gaza (Güney 2010). For these young men, 'Muslimness' had been converted into a political marker, in much the same way that Blackness had functioned for African-Caribbean youths in the 1980s.

Here, too, common feelings of sympathy and unity had arisen out of a collective distaste for the hegemonic activities of the 'white West' (i.e. the United States and Britain) that clearly superseded their shared religious attachment:

> *In fact, most of the youths I interviewed recognize suicide bombings as a final act of desperation and hopelessness rather than an act explicitly motivated by faith. Moreover, they consider the Palestinian suicide bombings in particular to be courageous acts by Muslims as a last means of resistance against the cycle of powerlessness, or even as a way of taking control. For these informants, the General Western critique of the (suicide) bombings and the mass media's portrayal of suicide bombers as 'Islamic fanatics' seem to be insignificant or merely symptomatic of the ignorance of the West. (Güney 2010)*

Abbas (2007) was dismayed to learn that six Muslim parliamentarians who wrote to the Prime Minister to express the feeling that British foreign policy was 'furthering the causes of militant suicide cults' quickly found themselves politically ostracized by their own party. The UK government's refusal to recognize any such link between 'home-grown terrorism' and foreign policy is further exemplified by their refusal to sanction any form of official inquiry. Abbas maintains that the continuation of a domestic policy based on the tightening up of anti-terrorist measures aimed primarily at Muslim communities, allied to an intensification of the 'war on terror' abroad, will inevitably reinforce the radicalization of British Muslim youths. This sentiment is echoed by Mythen and Walklate (2009), who concluded that, 'unless the issue of foreign and military policy can be openly debated – along with the shortcomings and errors therein – there is little likelihood that young Muslims will be supportive of a government that does not seem to understand them or listen to their opinions, let alone represent them'.

Conclusion

There can be no disputing that Beck's application of his Risk Society approach has been useful in helping to elucidate the relationship between global economic and technological developments and the corresponding spread of transnational terrorist activity. Beck's recent commentaries also eloquently attest to the ways in which the rhetorical and practical responses of Western liberal democracies, such as the UK and USA, have heightened public perceptions of the threat posed by terrorists (whether abroad or amidst their own societies) and unwittingly enhanced the power of those they are seeking to counteract or, better still, eliminate.

The evidence of this chapter does suggest, however, that Beck may have been guilty of overstating both the novelty and pervasiveness of the kind of terrorism occurring in the modern era. It also seems apparent that Beck does not adequately address, let alone explain, the underlying grievances that motivate and propel terrorist behaviour. Finally, although Beck alludes to the propensity of world leaders to consciously manipulate public perceptions of the terrorist threat, both to justify overseas incursions and help consolidate domestic cultures of control, it has generally been left to other scholars to show how such processes operate, and how they have invariably promoted an increased threat of terrorism by potential perpetrators residing inside and outside their own societies.

It may well be argued that, by focusing on the depiction of terrorism as a quintessentially 'aimless', 'unpredictable' and 'undifferentiating' component of the Risk Society, Beck has helped to distract us from the fact that terrorist atrocities tend, as they always have been, to be focused on those world powers whose foreign policies appear imperialistic, unjustifiable and irresistible to political reason or military opposition. Far from being an unfortunate or inevitable side effect of the growth of modernity, terrorism continues to be the last resort of people who feel they are being coerced into positions of cultural or economic subjugation. Such conditions are reversible, making it possible to envisage how British and American societies of future eras may find themselves feeling less endangered by the constant threat of terrorism. However, for this to happen will require their governments to go one step further than the more listening and inclusive 'cosmopolitan approach' advocated by Ulrich Beck. It will necessitate a drastic revision, or even abandonment, of current foreign and domestic policies.

References

Abbas, T. 2007. Muslim minorities in Britain: Integration, multiculturalism and radicalism in the post-7/7 period. *Journal of Intercultural Studies*, 28(3), 287–300.

Altheide, D.L. 2007. The mass media and terrorism. *Discourse and Communication*, 1(3), 287–308.

Beck, U. 1992. *Risk Society: Towards a New Modernity*. London: Sage.

Beck, U. 1998. Politics of risk society, in *The Politics of Risk Society*, edited by J. Franklin. Cambridge: Polity Press, 9–22.

Beck, U. 2002. The terrorist threat: World Risk Society revisited. *Theory, Culture and Society*, 19(4), 39–55.

Beck, U. 2006. Living in the world risk society. *Economy and Society*, 35(3), 329–45.

Beck, U. 2009. *World at Risk*. Cambridge: Polity Press.

Blair, T. 2006. *Speech on the Middle East to the Los Angeles World Affairs Council*. Available at: http://webarchive.nationalarchives.gov.uk/ [accessed: 19 August 2010].

Brighton, S. 2007. British Muslims, multiculturalism and UK foreign policy: 'Integration' and 'cohesion' in and beyond the state. *International Affairs*, 83(1), 1–17.

Croft, S. and Moore, C. 2010. The evolution of threat narratives in the age of terror: Understanding terrorist threats in Britain. *International Affairs*, 86(4), 821–35.

Duyvesteyn, I. 2004. How new is the New Terrorism? *Studies in Conflict and Terrorism*, 27, 439–54.

Fekete, L. 2004. Anti-Muslim racism and the European security state. *Race and Class*, 46(1), 3–29.

Frost, D. 2008. Islamaphobia: Examining causal links between the state and 'race hate' from 'below'. *International Journal of Sociology and Social Policy*, 28(11–12), 546–63.

Furedi. 2005. Terrorism and the politics of fear, in *Criminology*, edited by C. Hale, K. Hayward, A. Wahidin and E. Winkup. Oxford: Oxford University Press, 307–22.

Güney, U. 2010. 'We see our people suffering': The war, the mass media and the reproduction of Muslim identity among youth. *Media, War and Conflict*, 3(2), 168–81.

Hoffman, B. 2002. Rethinking terrorism and counterterrorism since 9/11. *Studies in Conflict and Terrorism*, 25, 303–16.

Kundnani, A. 2007. Integrationism: The politics of anti-Muslim racism. *Race and Class*, 48(4), 24–44.

Kundnani, A. 2009. *Spooked: How Not to Prevent Violent Extremism*. London: Institute of Race Relations.

Lambert, R. 2008. Salafi and Islamist Londoners: Stigmatised faith communities countering al-Qaida. *Crime, Law and Social Change*, 50, 73–89.

Manningham-Buller, D. 2010. *Evidence to The Iraq Inquiry*. Available at: http://www.iraqinquiry.org.uk/media/48331/20100720am-manningham-buller.pdf [accessed: 20 July 2010].

McEwan, I. 2001. Beyond belief. *The Guardian*, 12 September.

Mythen, G. and Walklate, S. 2006a. Criminology and terrorism: Which thesis? Risk Society or Governmentality? *British Journal of Criminology*, 46, 379–98.

Mythen, G. and Walklate, S. 2006b. Communicating the terrorist risk: Harnessing a culture of fear? *Crime, Media, Culture*, 2(2), 123–42.

Mythen, G. and Walklate, S. 2008. Terrorism – Risk and international security: The perils of asking 'What if?' *Security Dialogue*, 30(2–3), 221–42.

Mythen, G. and Walklate, S. 2009. 'I'm a Muslim, but I'm not a terrorist': Victimization, risky identities and the performance of safety. *British Journal of Criminology*, 49, 736–54.

Pantazis, C. and Pemberton, S. 2009. From the 'old' to the 'new' suspect community. *British Journal of Criminology*, 49, 646–66.

Sivanandan, A. 2006. Race, terror and civil society. *Race and Class*, 47(3), 1–8.

Spalek, B. and Lambert, R. 2009. Muslim communities, counter-terrorism and counter-radicalisation: A critically reflective approach to engagement. *International Journal of Law, Crime and Justice*, 36, 257–70.

Spencer, A. 2006. Questioning the concept of 'New Terrorism'. *Peace, Conflict and Development*, 8, 1–33.

Stohl, M. 2008. Old myths, new fantasies and the enduring realities of terrorism. *Critical Studies on Terrorism*, 1(1), 5–16.

Thomas, P. 2010. Failed and friendless: The UK's 'Preventing Violent Extremism' programme. *British Journal of Politics and International Relations*, 12, 442–58.

3

The Emergent Nature of Risk as a Product of 'Heterogeneous Engineering': A Relational Analysis of Oil and Gas Industry Safety Culture

Anthony J. Masys

Introduction

As systems become larger, more complex and more interdependent, new ways are required to identify latent risks. Actor-network theory (ANT) (Callon and Latour 1981, Latour 1987) provides a way of understanding the aetiology, operation and possible failure modes of complex socio-technical systems, like the NASA Space Transportation System ('Shuttle') vehicle, chemical plants and oil production platforms. Such systems emerge from a process of heterogeneous engineering – the purposeful assembly of technological, social, economic and political elements to meet some predetermined goal (extracting oil from under the ocean, for example). Dulac (2007) argues that complex socio-technical systems have a tendency to slowly drift from a safe state toward a higher-risk state, where they are highly vulnerable to small disturbances whereby seemingly inconsequential events can precipitate an accident. As a means of reducing the potential for large-scale disasters, and accidents associated with routine tasks, industry is now showing an increased interest in the concept of 'safety culture' (Cooper 2000). This chapter uses ANT to 'deconstruct' through a relational analysis, the oil and gas industry's safety culture.

Accidents such as the 1984 chemical release at Bhopal, the 1986 Chernobyl nuclear disaster, the 1986 Challenger Space Shuttle and 2003 Columbia Space Shuttle disasters, the 1988 Occidental Piper Alpha accident, the 1989 Exxon Valdez and 2010 BP Deepwater Horizon Oil Rig disaster in the Gulf of Mexico highlight the risks associated with operating complex socio-technical systems. According to Dörner (1996), the main challenges associated with understanding complex systems revolve around issues pertaining to the interdependency of factors, the complex nonlinear dynamic nature of the systems, and the uncertainty associated with the unobservable. Accidents involving socio-technical systems are therefore difficult to mitigate due to the inherent complexity of the system that stems from the interdependencies and relationality. Because of this, these systems are not well understood by those who design, manage and operate them (Chapman 2005). As such the inherent risk associated with these complex socio-technical systems '… arises not from a singular cause but from the interactions at the systemic level' and thereby necessitates the requirement to better understand the system interdependencies and dynamics (Miller 2009).

Germane to this work, the socio-technical system is a topic of inquiry within sociology that combines the social and technical paradigms and examines the relationship between them. As described by Coakes (2003), 'Socio-technical thinking is holistic in its essence; it is not the dichotomy implied by the name; it is an intertwining of human, organizational, technical and other facets'. Senge (1990) argues that since the world exhibits qualities of wholeness, the relevance of systemic thinking is captured within its paradigm of interdependency, complexity and wholeness. Although events can be considered to be discrete occurrences in time and space '… they are all interconnected. Events can be understood only by contemplating the whole' (Flood 1999).

Sociology offers an interesting approach for looking at the socio-technical elements of complex systems through the application of ANT. The systems perspective of ANT looks at the inter-connectedness of the heterogeneous elements characterized by the technological and non-technological (human, social, organizational) elements thereby defining a web of relations (network space). Yeung (2002) notes that much of the work that draws on ANT places its analytical focus on unearthing the complex web of relations between humans and non-humans. The interaction of non-human actors with the human actors gives shape and definition to identity and action. Latour (1994) argues that '… it is impossible even to conceive of an artefact that does not incorporate social relations, or to define a social structure without the integration of non-humans into it. Every human interaction is socio-technical'. The 'social' is thereby described as 'materially heterogeneous' (Callon and Law 1997). The theoretical perspective of ANT challenges the fractured view of the world that stems from dualistic thinking, as described in Murdoch (1997). The holistic perspective of ANT makes it well suited to facilitate an examination of the complex socio-technical systems associated with accident aetiology, and in particular an analysis of safety culture.

Safety Culture

The term 'safety culture' first emerged in 1987 following an investigation of the 1986 Chernobyl disaster (Cooper 2000). Since the appearance of the term, many definitions have emerged from the safety literature (Cox and Cox 1991, Mearns *et al.* 1998, Cooper 2000, Mearns, Whitaker and Flin 2003). Within the application domain of the oil and gas industry, Mearns *et al.* (1998) define safety culture as 'the attitudes, values, norms and beliefs which a particular group of people share with respect to risk and safety'. This definition, like many within the literature, links safety culture to the 'human' actor. Toft and Reynolds (1999) argue that safety culture can be defined as those sets of norms, roles, beliefs, attitudes and social and technical practices within an organization which are concerned with minimizing the exposure of individuals to conditions considered to be dangerous. This recognizes the important influence that physical and management risk control measures have on an organization's safety culture.

Recognizing the social and technical components of safety, the socio-technical model of safety culture argues for the proactive integration of safety into organizational structures and processes and the requirement for the joint optimization of technology and work organization taking into account both material and immaterial characteristics of an organization (Grote and Künzler 2000). Based on the socio-technical systems approach, an understanding of safety culture finds itself deeply rooted in assumptions about the interplay of people, technology, and organizations in their relation to safety. In addition it recognizes shared as well as conflicting norms, practices that permeate the socio-technical system and exist within the human, physical and informational domains (Grote and Künzler 2000). This perspective thereby supports the application of the actor network worldview facilitating a relational analysis of the safety culture within the socio-technical system.

Risk

In a complex system, actions can have unanticipated consequences, and feedback effects can produce nonlinear outcomes. Normal accidents are those situations where unanticipated interactions, often involving multiple component failures, occur. Perrow (1984) argues that our ability to manage high-risk technology is insufficient to prevent accidents and crises. Such accidents are outcomes of inscribed and inherent features of complex technologies. In a sense, complex technologies engender risks that emerge or are triggered through unanticipated interactions among elements of systems.

Modern approaches to risk, crisis and disaster management give pre-eminence to expert knowledge and the formal state apparatus of disaster response and recovery. Late-modern approaches, in contrast, recognize lay knowledge and the benefits of public participation in disaster response and recovery. In late modernity, risk refers to that which cannot be known — to unquantifiable uncertainties: as described by Gephart, Van Maanen and Oberlechner (2009), 'These uncertainties have undermined the role of experts to define risks and their acceptability, and have led to the politicization of risk conflicts'. Complex socio-technical systems are risky because their operation and performance is dependent upon many interdependent factors (Perrow 1984, Dörner 1996). Given the inherent interdependencies and complexity of a constructed socio-technical system, the resulting challenges associated with risk assessment and management also emerge. Late-modern approaches to risk, crisis and disaster management both recognize and exploit complexity, specifically the existence of different worldviews (reality constructs). Resonating with late-modernity, Perrow argues that the sharing and questioning of mental models across strata can help improve safety. In congruence with the late-modern approach to risk management, ANT with its inherent multi-vocality facilitated by the relational analysis, provides a suitable lens for examining risk and safety culture.

Actor Network Theory: Understanding the Socio-technical System

Systems thinking, according to Senge (1990), 'is a discipline for seeing wholes. It is a framework for seeing interrelationships rather than things, for seeing patterns of change rather than static snapshots'. As a worldview, systems thinking recognizes that systems cannot be addressed through a reductionist approach that reduces the systems to their components. The behaviour of the system is a result of the interaction and interrelationships that exist, thereby acknowledging emergent behaviours and unintended consequences. Supporting this Dekker (2006) argues '... it is critical to capture the relational dynamics and longer term socio-organizational trends behind system failure' thereby supporting a systemic perspective of accident aetiology.

ANT is a theoretical perspective that has evolved to address the socio-technical domain and in particular the conceptualization of the 'social'. This perspective challenges the way we think of agency, the human and non-human. The application of the ANT perspective (terms and concepts) has been instrumental in revealing insights within such fields as information technology, organizational theory, geography, medical anthropology and psychology. Latour (2005) introduces ANT as a 'relativistic perspective' that challenges the current paradigm associated with the sociology of the social. It is through this examination of the 'social' that the inherent complexity associated with understanding accident aetiology is revealed.

ANT treats both human and machine (non-human) elements in a symmetrical manner, thereby facilitating the examination of a situation (like an accident) where, Callon (1999) argues, '... it is difficult to separate humans and non-humans', and where 'the actors have variable forms and competencies'. As noted by Ashmore, Woolfitt and Harding (1994), through ANT:

> ... the assumption of the ontological primacy of humans in social research and theory is suspended. Non-human entities, traditionally overlooked in sociological accounts of the social world, take their rightful place as fully fledged actants in associations, relations, and networks.

Fundamental concepts within ANT are the conceptualization of the actor and the network. An actor-network, as described by Latour (1987) and Callon (1986a, 1991), is characterized as a network that is inherently heterogeneous, where the relations between the actors are important, rather than their essential or inherent features. The actor, whether technical or non-technical, is examined within the context of a heterogeneous network. In fact the actor is a network in itself '... in the same way, elements in a network are not defined only by their "internal" aspects, but rather by their relationships to other elements, i.e., as a network' (Aanestad and Hanseth 2000). The actors or actants of ANT can be humans, organizations, cultures, ideas, animals, plants or inanimate objects and are described in terms of the alliances and exchanges they exhibit in the interconnected network of relations. Latour (1987) defines the word network in terms of connections of scattered resources that coalesce as a net that seem to extend everywhere. The heterogeneous elements that make up the network are not fixed but are defined in relation to the other elements in the system. Williams-Jones and Graham (2003) argue that:

> *ANT is an approach that is interested in the tensions between actor, network and technology, and how they manifest in practice (Law 1999). Failed networks are thus often a fruitful place for study, because it is here that the actor-networks reveal themselves and the norms and values built into technologies are made apparent.*

The lens of ANT facilitates the view of the world in terms of heterogeneous elements, thereby demonstrating that nature and society are 'outcomes' rather than 'causes' emerging from a complex set of relations (Murdoch 1997).

Fundamental processes within ANT are inscription and translation. Inscription refers to the way technical artefacts embody patterns of use: technical objects thus simultaneously embody and measure a set of relations between heterogeneous elements (Akrich 1992). Monteiro (2000) argues that the notion of inscription may be used to '... describe how concrete anticipations and restrictions of future patterns of use are involved in the development and use of a technology'. Inscriptions enable action at a distance by creating 'technical artefacts' that ensure the establishment of an actor's interests such that they can travel across space and time and thereby influence other work (Latour 1987). Inscripted artefacts such as texts and images are central to knowledge work (Wickramasinghe *et al.* 2007) and thereby can shape sensemaking, decision-making and action.

The process of translation has been described as pivotal in any analysis of how different elements in an actor network interact (Somerville 1997). As a transformative process, translation emphasizes '... the continuous displacements, alignments and transformations occurring in the actor network' (Visue 2005). Translation rests on the idea that actors within a network will try to enrol (manipulate or force) the other actors into positions that suit their purposes. When an actor's strategy is successful and it has organized other actors for its own benefit, it can be said to have translated them. The process of translation comprises undertones of power mechanisms as described in Callon (1986a). Hernes (2005) argues that translation can be regarded as '... negotiations, intrigues, calculations, acts of persuasion and violence, thanks to which an actor or force takes, or causes to be conferred on itself, authority to speak or act on behalf of another actor or force'. The actor network is therefore comprised of human and non-human elements through a series of negotiations such that actors seek to impose definitions of the situation on others (Callon 1986b).

Within the context of this chapter through the application of ANT, complex systems are seen as heterogeneous actor-networks that consist of a particular configuration of more or less aligned human and non-human components. Examination of actors such as those characterized as technologies or policies, facilitates an exploration of how these 'actors' mediate action and how they are entangled in local techno-social configurations. The three elements from ANT that are particularly relevant to the study of safety culture in the oil and gas industry include the principal of symmetry; the focus on actor-networks and dissolving dualisms; and the emphasis on processes of translation (Van der Duim 2005).

Methodology

Rooted in ethnography, ANT seeks to understand relational dynamics from the inside-out, by following the actors and asking how the world looks through the eyes of the actor doing the work. Through this approach emerge issues pertaining to the roles that tools and other artefacts (actants) play in the actor network in the accomplishment of their tasks (Dekker and Nyce 2004). The foundational method associated with an ANT study, as suggested by Callon (1986a) and Latour (1996, 2005), is to 'follow the actors' and 'let them set the framework and limits of the study themselves' (Tatnall 2000). Through this approach we search out the interactions, negotiations, alliances and networks that characterize the network space. The relational mapping associated with ANT resonates with the propositional networks described in detail in Stanton *et al.* (2009). Latour (1987) remarks:

> ... we have to be as undecided as possible on which elements will be tied together, on when they will start to have a common fate, on which interest will eventually win out over which. In other words, we have to be as undecided as the actors we follow ... The question for us, as well as those we follow, is only this: which of these links will hold and which will break apart?

With regards to the question of the oil and gas industry and safety culture, we leverage the comments from Harbers (2005) who argues that '... we are confronted here with a hybrid situation in which human beings and technology are tightly interwoven – a mixture, a muddle of man and machine'. We address this issue through the concept of the 'hybrid collectif' that emerges from the analysis.

Case Studies

Case studies from the Oil and Gas industry provide the context of the analysis of safety culture eliciting, through the lens of ANT, insights into the problem space of risk within complex socio-technical systems. The three case studies that inform the ANT examination of safety culture include the 1988 Occidental Piper Alpha accident, 2003 Shell Brent Bravo accident and the 2010 BP Deepwater Horizon accident.

PIPER ALPHA

The Cullen Report (1990) describes an accident aetiology that arises from a complex socio-technical system. The 1988 Piper Alpha accident represents one of the most catastrophic events to occur within the oil and gas industry. The fatal accident resulted in the death of 167 workers. The cause of the disaster is associated with a failure in the permit-to-work system which caused a breakdown in communications between the day shift and the night shift. This led to the use of machinery which was undergoing maintenance and caused the escape of gas from an insecurely fastened temporary flange. Throughout the evolution of the accident a series of failures and errors of judgment contributed to the overall scale of the disaster.

The Cullen Report (1990) was highly critical of the management system in the company. It was noted that managers had minimal qualifications, which resulted in poor practices and ineffective audits. The report highlighted issues pertaining to regulatory control of offshore installations; adherence to standard operating procedures associated with the permit-to-work system; quality of safety management; need for safety training; auditing practices; and fire and explosion protection as implicated in the accident aetiology. As noted in Lavelette and Wright (1991), extensive sub-contracting characterized the workforce offshore. Of the employees on Piper Alpha on the night of the accident, only 17 per cent were Occidental (primarily management). This sub-contracting system enabled Occidental to push some of the costs of oil production onto the sub-contracting firm.

Occidental management was criticized for not being prepared for a major emergency. This stemmed from Occidental's adoption of a '... superficial attitude to the risk of major hazards'. Their failure to put in place appropriate safety systems made them incapable of responding effectively to foreseeable dangers. Processes and procedures like the permit-to-work (PTW) process and handover processes were habitually departed from. Additionally the Cullen Report identifies significant shortcomings within the Department of Energy with regards to the conduct of safety inspections of the platform. In Miller (1991) the inspections were characterized as '... superficial to the point of being of little use as a test of safety on the platform' (Cullen 1990).

BRENT BRAVO

Following the Cullen Report, corrective action within the industry was taken to address the over 100 recommendations. Unfortunately in 2003 an accident claimed the lives of two oil rig workers on the Brent Bravo offshore platform. The investigation highlighted defects in the system of working which contributed to the accident. It was reported that lapses on Brent Bravo were similar to the kind of wider problems identified in the Platform Safety Management Review (PSMR) conducted in 1999 by Former Senior Manager Bill Campbell (Macalister 2006). The 1999 PSMR highlighted how equipment was being operated in a dangerous condition. His report alleged that '... platform managers reported 96% compliance with safety critical maintenance while the actual levels of compliance were 14%' (BBC News 2006).

The two fatalities associated with the Brent Bravo accident of 11 September 2003 resulted from the inhalation of hydrocarbon vapours. The cause of the accident stemmed from the release, and vaporization, of liquid hydrocarbons through a hole, caused by corrosion, in the closed drain degasser rundown line within the utility shaft of the Brent Bravo offshore platform. The 1999 audit of the Brent Bravo and 2003 accident investigation both recognize the existence of inappropriate attitudes, skills and behaviour on Brent Bravo that contributed to the fatalities. In particular inadequate communication, ineffective management controls, inappropriate maintenance practices resulting in defective physical systems and failure to manage and conduct appropriate risk assessments are highlighted as affecting the integrity of the system.

DEEPWATER HORIZON

The 2010 oil spill in the Gulf of Mexico is now considered the largest offshore spill in US history. The spill stemmed from a sea-floor oil gusher that resulted from the 20 April 2010 Deepwater Horizon drilling rig explosion. The explosion killed 11 platform workers and injured 17 others. The Interim Report (2010) argues that the incident was precipitated by the decision to proceed to temporary abandonment of the exploratory well despite indications from several repeated tests of well integrity that the cementing processes failed to provide an effective barrier to hydrocarbon flow.

A preliminary examination of the evidence (Bea 2010) suggests that the disaster was preventable had existing progressive guidelines and practices been followed. Some of these guidelines are implemented internationally where the same industry players, including BP, operate. Moreover, other existing US guidelines that were waived by the responsible regulatory authority could have prevented this incident (Bea 2010). The accident aetiology stems from the failure of a number of processes, systems and equipment. There were multiple control mechanisms – procedures and equipment – in place that should have prevented the accident or reduced the impact of the spill. Preliminary evidence collected (Bea 2010) suggests: failures in the effective evaluation of the likelihood and consequences of malfunctions; failure to adhere to established standards; improper well design; faulty quality assurance and quality control; ineffective decision making regarding deep water drilling practices; ineffective situation awareness which resulted in warning signs not being detected, analysed or corrected; improper operating procedures; flawed design of safety defences. Of particular note are the records which clearly show excessive economic and schedule pressures which resulted in compromises in the quality and reliability of the complex socio-technical system. The various failures described in the Interim Report (2010) identify the lack of a suitable approach for anticipating and managing the inherent risks, uncertainties and dangers associated with deepwater drilling operations and a failure to learn from previous near-misses. Of particular concern was the apparent lack of a systems approach that would integrate the multiplicity of factors potentially affecting the safety of the well, and inform and assess operational decisions from perspectives of well integrity and safety.

The common thread that connects all three accidents is the nature of the accident aetiology that is characterized by the inherent interconnectivity and interdependence of human and non-human actors. It points to a culture of complacency rooted in illusions of certainty, thereby affecting the sensing of weak signals and organizational learning.

Discussion

The ANT-informed analysis of the socio-technical systems characterized by the case studies reveals how latent errors/resident pathogens contribute to failure modes. Following the actor traces a complex relational network (described by Callon and Law (1995) as the *hybrid collectif*). It is within this hybrid collectif that interdependencies and complexities (of the latent errors/resident pathogens) emerge. Dekker (2005) views the:

> ... *socio-technical system not as a structure consisting of constituent departments, blunt ends and sharp ends, deficiencies and flaws, but as a complex web of dynamic, evolving relationships and transactions.*

The simple cause and effect paradigm does not account for the interdependent nature of the system. Rather, emergent properties are only visible when viewing the problem space from a systems perspective. This necessitates that we think in terms of '... processes of change rather than snapshots' (Senge 1990). This systems perspective (that is realized by ANT) reveals the interdependent actor network (human or non-human) thereby shifting our awareness from linear causality to complex causality. By tracing the relational actor network that describes the accident aetiology, safety culture emerges as a construct resident within a network of heterogeneous elements that shapes, and is shaped by, inscription and translation processes. Safety culture is not just something that exists between people: objects are viewed as not passive; they are actors that actively shape and transform that which they are part of (Viseu 2005). What emerges from the analysis of all the case studies is that '... Action, intentionality ... derive from relations between entities rather than from either individuals or totalities' (Lockie 2004). Hence safety culture derives from a relational network.

The interdependencies between the human, technical and informational are captured by Lavelette and Wright (1991) who argue that:

> *The picture presented in the evidence to the Inquiry is of a management system offshore whose priority is to maintain production and get oil onshore as quickly as possible. These priorities bring with them a high cost in terms of human safety. When machines and equipment go wrong in the North Sea, there is often not the manpower, facilities or time to carry out official procedures without directly affecting production. As a result, ad hoc 'pressure management' decisions are taken. Hence, the*

> *unusual and ad hoc become routinised, normalised and accepted within the context of the offshore work environment.*

Similarly, the failures and missed indications of hazards that preceded the Deepwater accident (such as the results of repeated negative-pressure tests) suggest an insufficient consideration of risk and a lack of operating discipline. As described in the Interim Report (2010), the decisions regarding risk also raise questions about the adequacy of operating knowledge on the part of key personnel. The net effect of these decisions was to reduce the available margin of safety (that took into account the complexities of the hydrocarbon reservoirs and well geology, and the subsequent changes in the execution of the well plan).

The accident aetiology emerges as a function of policies, organizational management, technical and informational factors (Figure 3.1). For example, the PTW system both within the Piper Alpha and Brent Bravo cases is designed to facilitate senior platform management approval and situation awareness of the status of all ongoing work. Lavelette and Wright (1991) argue that 'the evidence presented at the Piper Alpha inquiry makes it clear that this system was hopelessly inadequate'. Coupled with that, the handover procedure used to support the PTW was dysfunctional and represented 'the complete breakdown of communication between all the Control Room staff on the day and night shifts' (Lavelette and Wright 1991).

This resonates with Vaughan's (1996) analysis of the Challenger accident in which she states that 'Norms – cultural beliefs and conventions originating in the environment – create unreflective, routine, taken-for-granted scripts that become part of the individual worldview. Invisible and unacknowledged rules for behaviour, they penetrate the organization as categories of structure, thought and action that shape choice in some directions rather than others'. Vaughan (1996) describes the normalization of deviance as something common to organizations and individuals that can result in mistake, mishap and misconduct. Applicable to the oil and gas industry, the normalization of deviance forces one to recognize the social construction of mistake. Vaughan (1996) states that the normalization of deviance '… reminds us of the power of political elites, environmental contingencies, history, organizational structure, and culture as well as the impact of incrementalism, routines, information flows and taken-for-granted assumptions in shaping choice in organizations'.

Conflicting goals and policies lead to the adoption of work practices where safety considerations have a relatively low priority. Pertaining to the Deepwater

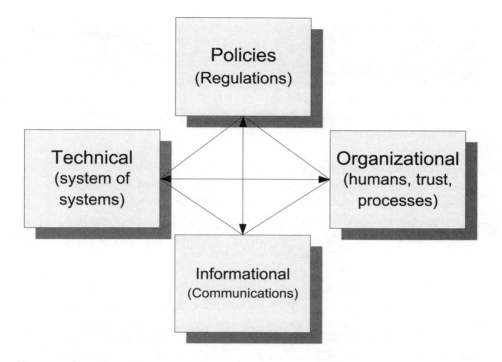

Figure 3.1 **Socio-technical system view of interdependencies contributing to accident aetiology**

incident, available evidence suggests there were insufficient checks and balances for decisions involving both the schedule to complete well abandonment procedures and considerations for well safety. As with Piper Alpha, lack of safety offshore was therefore built into offshore work practice (Lavelette and Wright 1991). In other words, pressures and goals inscribed into the system translated processes, people and technologies to support production goals that created conflicts with safety goals. The superficial inspection of Piper Alpha and the inadequately trained, guided and led inspections (Miller 1991) highlight how human, informational and organizational interdependencies converge to become hardwired into the system, thereby creating illusions of certainty (Masys 2010). The emergent safety culture reflects these dysfunctionalities (associated with inscribed assumptions, beliefs and values about safety within the human, physical and informational domains).

Latent errors/resident pathogens emerge within the system as flaws in design guidelines and design practices, misguided priorities in management and conflicting priorities. Of note is that Occidental Petroleum's management had been warned earlier that the platform could not survive prolonged exposure to high-intensity fire. The warning, however, was ignored because the event was judged too unlikely to be taken seriously. The sensemaking associated with the decision was based on an error in reasoning and, apparently, the incorrect assumption of independence in the successive failure events (Pate-Cornell 1993) thereby revealing inappropriate mental models of risk and safety. Similarly, in the Deepwater Horizon case:

> The decision to accept the results of the negative-pressure test as satisfactory – rationalized as being the result of some hypothesized 'bladder effect' (or annular compression) – without review by adequately trained shore-based engineering or management personnel suggests a lack of onboard expertise and of clearly defined responsibilities and the associated limitations of authority. (Interim Report 2010)

The mental models of safety were reflected and inscribed within the system of systems (that comprised both the social and technical domains as reflected in the '… superficial attitude to the assessment of the risk of major hazards' (Pate-Cornell 1993)). In a similar manner, the Brent Bravo 1999 Audit Report and 2003 investigation report highlighted the existence of inappropriate attitudes, skills and behaviour with regards to safety. These were shaped by the policies and conflicting goals that were inscribed into the socio-technical system. Vaughan (1996) argues '… how small changes – new behaviours that were slight deviations from the normal course of events – gradually [become] the norm, providing a basis for accepting additional deviance'.

By following the actors and mapping the interrelationships, three intersecting domains emerge that characterize safety culture: physical, human and informational. The intersection of these domains represents the hybrid collectif (represented by Φ).

ANT analysis reveals that:

> … if one were to try to draw a map of all of the actors present in any interaction, at any particular moment in time, instead of a well-demarcated frame, one would produce a highly convoluted network with a multiplicity of diverse dates, places and people. (Dolwick 2009)

The accident aetiology is therefore not an instant in time but the entanglement of an actor network, of multiple spaces and multiple times. Safety culture emerges

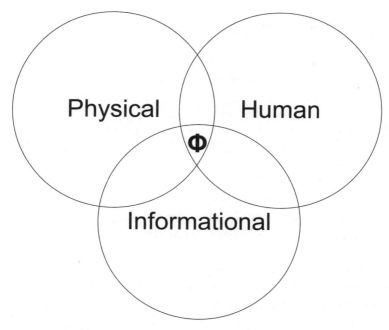

Figure 3.2   **Hybrid collectif**

as an entanglement whereby the actors (human, physical and informational) are relationally linked with one another in webs or networks. These actors, due to their inherent interdependency, make a difference to each other. They make each other be (Dolwick 2009). People, organizations, technologies and politics are the result of heterogeneous networks (Cressman 2009). Safety culture is not something abstract that resides within the minds of people but is rather:

> ... *an emerging property of a socio-technical system, the final result of a collective process of construction, a "doing" that involves people, technologies and textual and symbolic forms assembled within a system of material relations. (Gherardi and Nicolini 2000)*

In this sense safety culture is a product of heterogeneous engineering.

Drawing upon the socio-technical model of safety culture, and viewed through the lens of ANT, it is understood that:

1.   Safety culture is not an abstract mental model, but is a construct resident within a network of heterogeneous elements.

2.   Non-humans have significance and are not simply resources or constraints

    a.   Non-humans intervene actively to push action in unexpected directions.

3.   Action results from the complex interactions resident within the actor network that is dynamically shaped by inscription and translation processes.

4.   The actor network lens reveals that action cannot be explained in a reductionist manner, as a firm consequence of any particular previous action. (Callon and Law 1997)

Illusions of certainty (Masys 2010) emerge from the analysis illustrating how hardwired politics became inscribed within the system and shaped sensemaking, decision-making and action. For example, the oil production design capability of Piper Alpha was far exceeded, thereby resulting in an erosion of safety features (Pate-Cornell 1993). With a safety culture that values production over safety, there exists insufficient feedback in the system to sense weak signals. The disaster had a long incubation period characterized by a number of discrepant events signalling potential danger. Practical drift (Snook 2000) helps explain how safety culture can become dysfunctional:

> Practical drift is the slow, steady uncoupling of local practice from written procedure. It is this structural tendency for subunits to drift away from globally synchronized rule-based logics of action toward locally determined task-based procedures that places complex organizations at risk.

This is reminiscent of Vaughan's (1996) analysis:

> [B]y linking environment, organization and decision making, the [Challenger] case alerts us to the subtlety of definitional processes, affirming the power of historic actions in organizations that become solidified into norms, standard operating procedures and a shared worldview that shapes future choices.

This emerges from the translation processes where policies shape not only processes and procedures but also become inscribed within the physical infrastructure and common practices on the oil rig. This is consistent with Turner and Pidgeon (1997) who suggest that:

> disaster-provoking events tend to accumulate because they have been overlooked or misinterpreted as a result of false assumptions, poor communications, cultural lag and a misplaced optimism.

This resonates with Snook's (2000) theory of practical drift.

Safety Culture and Heterogeneous Engineering

Members of a Deepwater Horizon federal investigative panel 'blasted' BP for apparently failing to improve its safety culture after a string of accidents over the last decade (Hammer 2010). Events like the Deepwater Horizon accident, which killed 11 rig workers, the 2005 refinery explosions at BP's Texas City, Texas, facility that killed 15 employees, the near capsizing of BP and ExxonMobil's Thunder Horse rig off the Louisiana coast in 2005 and two near-blowouts of shallow-water wells in 2002, show a pattern of safety culture that is not working. Hammer (2010) reports how 'after the Texas City incident, the company spent $1.4 billion to "change the mechanical integrity" of the buildings that are close to refineries or to move them ... the other incidents also led to equipment and procedural changes'. This illustrates BP's failure to view the problem from a systems perspective. Resorting to a default techno-centric view obscured the fact that the problem resided within the company's safety culture (Hammer 2010). Failure to understand the reverberations of technological or process change on the operational system and associated safety culture hinders the understanding of important issues surrounding the evolution of accidents:

> ... Technologies are not simply passive and are never value neutral .... During the design phase, objects have embedded within them a 'script' or set of instructions that determine how the technology will function and the extent to which it may be shaped by other actors. (Williams-Jones and Graham 2003)

Rosen and Rappert (1999) argue that the '... design of artefacts can prohibit certain uses or compel particular kinds of uses' and become reflected in the design and implementation of safety practices and supporting architecture and policies that shape decision making. Turner (1976) argues that many accidents result from 'failures of foresight'. He describes this as the 'accumulation of an unnoticed set of events which are at odds with the accepted beliefs about hazards and the norms for their avoidance'. Reason (2004) points out that 'the path to adverse incidents is paved with false assumptions'. These scripts reflect and affect the safety culture. As noted in Faraj *et al.* (2004) '... beliefs can arise either from the technology histories of particular actors or from interdependent relationships among multiple actors'. Safety emerges as a property of a socio-technical system, evolving through dynamic and nonlinear interactions between actors (that involve people, technologies and textual and symbolic forms assembled within a system of material relations). What emerges from the analysis of the Piper Alpha, Brent Bravo and Deepwater Horizon disasters is that safety, in the oil and gas industry, does not have anything like the priority it should have.

Effective risk management reflects a safety culture that recognizes 'near-misses' as opportunities to improve, thereby facilitating organizational learning. For example on 8 March 2010, a month before the accident, BP discovered that drilling fluid had leaked into rocks 5,000 metres below the sea floor instead of returning to the surface via a pipe. The leak caused the pressure in the well to drop, allowing oil and gas to flow into the well bore. The Interim Report (2010) highlights how this near-miss should have alerted companies to the imminent danger and need for more detailed tests. Instead, BP pressed on with its efforts to drill and seal the well, ready for the commercial extraction of oil and gas (Gupta 2010). The various failures described in the report indicate the lack of a suitable approach for anticipating and managing inherent risks, uncertainties and dangers and failure to learn from near-misses.

Hughes (2010) describes how 'officials investigating causes of the Gulf of Mexico oil spill suggested that the disaster wasn't an anomaly but reflected systemic problems in the offshore-drilling industry – a conclusion that could have repercussions for the entire oil industry ... it is a systemic problem'. Creating a safety culture that embraces foresight requires an understanding of the interdependencies that reside within the actor network. Heterogeneous engineering recognizes the need for proactive risk management, in which multi-vocality reveals risks and their interrelationships in a proactive manner. This approach embraces the central tenets of systems thinking. It is holistic in its approach and facilitates a continuous process of monitoring all identified hazards; predicting risk scenarios; and using organizational learning to profit from past mistakes. Heterogeneous engineering facilitates the challenging of mental models, assumptions, beliefs and values and corrects two habitual shortcomings – a failure of foresight (Toft and Reynolds 1999) and a failure to learn. Heterogeneous engineering works at the level of the hybrid collectif: it recognizes the interdependencies that exist within the physical, human and informational domains. Through heterogeneous engineering, sources of complexity in socio-technical systems are revealed. This knowledge should guide the process of identifying potential hazards and enable designers, managers and operators to become more vigilant and reflexive (Chapman 2005).

Challenging the Mental Models of Risk Production

Mental models are '… deeply ingrained assumptions, generalizations, or even pictures or images that influence how we understand the world and how we take action' (Senge 1990). Chapman and Ferfolja (2001) discussed several processes through which mental models become flawed in industrial settings (resulting in the misreading of situations pertinent to the oil and gas industry's safety culture). These processes include '… retaining outdated knowledge that no longer applies, accepting unreliable sources of information at face value, and missing out on critical data because of poor communication within the work organization' (Chapman 2005). Illusions of certainty have everything to do with expectations. As Weick and Sutcliffe (2007) argue:

> … Expectations are built into organizational roles, routines, and strategies. These expectations create the orderliness and predictability … Expectations, however, are a mixed blessing because they create blind spots. Blind spots sometimes take the form of belated recognition of unexpected, threatening events. And frequently blind spots get larger simply because we do a biased search for evidence that confirms the accuracy of our original expectations.

The case studies support Woods's (2003) observation that:

> An organization usually is unable to change its model of itself unless and until overwhelming evidence accumulates that demands revising the model. This is a guarantee that the organization will tend to learn late, that is, revise its model of risk only after serious events occur.

Heterogeneous engineering challenges this model to suggest a proactive approach to managing risk and being sensitive to weak signals. As an emergent property, safety is dynamic and contextual. Safety requires constant vigilance.

The erosion of safety culture stems from a failure to recognize the interdependencies and latent errors/resident pathogens that reside within the actor network. A reductionist approach to accident investigation (White 1995) fails to capture this aspect of complex socio-technical systems. A system perspective of the problem space is essential.

Conclusion

The Interim Report (2010) describes the engineering and drilling operations associated with drilling offshore, especially in deep water, as exceedingly complex, in that they involve a wide range of technologies and a large number of contractors. Behind the technical cause of the disaster rests a series of routinized, *ad hoc* managerial decisions and 'bad practices'. The disaster on Piper Alpha cannot be understood without this organizational context (Lavelette and Wright 1991). To manage risks involving socio-technical systems requires an approach that focuses on studying the synergistic effects of a system's actors.

Relational analysis identifies safety culture as emerging from the hybrid collectif – from the physical, informational and human domains. What the case studies show through this relational analysis is how the inscription and translation processes within ANT shape action, perception and decision-making. Latent errors/resident pathogens emerge from the analysis. Safety culture is sensitive to the interdependencies of the socio-technical system: '[T]he world is not a fixed stage independent of the actors who pass through it. The actors are participants whose actions contribute to the creation of the stage' (Flach *et al.* 2008). Safety culture emerges from the network of heterogeneous elements that defines the actor network. It represents the interdependency associated with material and immaterial reality. Heterogeneous engineering promotes and requires broader participation and discussion among various groups and stakeholders. It embraces the late-modern approach to risk analysis and management by leveraging alternative views of the problem space.

References

Aanestad, M. and Hanseth, O. 2000. Implementing Open Network Technologies in Complex Work Practices: A Case from Telemedicine. Proceedings from the IFIP WG 8.2 conference IS2000, 10–12 June. Aalborg, Denmark:

Organizational and Social Perspectives on Information Technology. Kluwer Academic Publishers, 355–69.

Akrich, M. 1992. The De-Scription of Technical Objects, in *Shaping Technology, Building Society: Studies in Sociotechnical Change*, edited by W. Bijker and J. Law. Cambridge, MA: MIT Press, 205–24.

Ashmore, R., Woolfitt, R. and Harding, S. 1994. Humans and others, agents and things. *American Behavioural Scientist*, **37**(6), 733–40.

BBC News 14 June 2006. Shell 'ignored accident warning'. Available at: http://news.bbc.co.uk/2/hi/5077886.stm [accessed: 4 January 2010].

Bea, R. 2010. Failures in the Deepwater Horizon Semi-Submersible Drilling Unit, Centre for Catastrophic Risk Management: Department of Civil and Environmental Engineering. Berkeley: University of California.

Callon, M. 1986a. The Sociology of an Actor-Network: The Case of the Electric Vehicle, in *Mapping the Dynamics of Science and Technology*, edited by M. Callon, J. Law and A. Rip. London: MacMillan Press, 19–34.

Callon, M. 1986b. Some Elements of a Sociology of Translation: Domestication of the Scallops and the Fishermen of St Brieuc Bay, in *Power, Action and Belief: A New Sociology of Knowledge?* edited by J. Law. London: Routledge and Kegan Paul, 196–233.

Callon, M. 1991. Techno-economic Networks and Irreversibility, in *A Sociology of Monsters? Essays on Power, Technology and Domination, Sociological Review Monograph*, edited by J. Law. London: Routledge, 132–61.

Callon, M. 1999. Actor-Network Theory: The Market Test, in *Actor Network and After*, edited by J. Law and J. Hassard. Oxford and Keele: Blackwell and the *Sociological Review*, 181–95.

Callon, M. and Latour, B. 1981. Unscrewing the Big Leviathan: How Actors Macro-structure Reality and How Sociologists Help Them To Do So, in *Advances in Social Theory and Methodology: Towards an Integration of Micro and Macro Sociologies*, edited by K. Knorr-Cetina and A.V. Cicourel. Boston, MA: Routledge and Kegan Paul, 277–303.

Callon, M. and Law, J. 1995. Agency and the hybrid collectif. *The South Atlantic Quarterly*, **94**(2), 481–507.

Callon, M. and Law, J. 1997. After the individual in society: Lessons on collectivity from science, technology and society. *Canadian Journal of Sociology*, **22**(2), 165–82.

Chapman, J. 2005. Predicting technological disasters: Mission impossible? *Disaster Prevention and Management*, **14**(3), 343–52.

Chapman, J.A. and Ferfolja, T. 2001. Fatal flaws: The acquisition of imperfect mental models and their use in hazardous situations. *Journal of Intellectual Capital*, **2**(4), 398–409.

Coakes, E. 2003. Socio-technical Thinking – An Holistic Viewpoint, in *Socio-technical and Human Cognition Elements of Information Systems*, edited by S. Clarke, E. Coakes, M.G. Hunter, and A. Wenn. Hershey: Information Science Publishing, 1–4.

Cooper, M.D. 2000. Towards a model of safety culture. *Safety Science*, **36**, 111–36.

Cox, S. and Cox, T. 1991. The structure of employee attitudes to safety: A European example. *Work and Stress*, **5**, 93–106.

Cressman, D. 2009. *A Brief Overview of Actor-network Theory: Punctualization, Heterogeneous Engineering and Translation*, 1–17. Available at: http://www.sfu.ca/cprost/docs/A%20Brief%20Overview%20of%20ANT.pdf [accessed: 2 April 2009].

Cullen, W.D. 1990. *The Public Inquiry into the Piper Alpha Disaster*. London: HMSO.

Dekker, S. 2005. *Why We Need New Accident Models*, Technical Report 2005–02, Lund University School of Aviation. Available at http://www.lu.se/upload/Trafikflyghogskolan/TR2005-02_NewAccidentModels.pdf [accessed: 26 October 2009].

Dekker, S. 2006. Resilience Engineering: Chronicling the Emergence of Confused Consensus, in *Resilience Engineering: Concepts and Precepts*, edited by E. Hollnagel, D.D. Woods, and N. Leveson. Aldershot: Ashgate Publishing Ltd., 77–92.

Dekker, S. and Nyce, J.M. 2004. How can ergonomics influence design? Moving from research findings to future systems. *Ergonomics*, **47**(15), 1624–39.

Demarest, M. 1997. Knowledge management: an introduction. Available at: http://www.noumenal.com/marc/km1.pdf [accessed: 1 May 2008].

Dolwick, J.S. 2009. 'The social' and beyond: Introducing actor-network theory. *Journal of Maritime Archaeology*, **4**(1), 21–49.

Dörner, D. 1996. *The Logic of Failure: Recognizing and Avoiding Error in Complex Situations*. Cambridge, MA: Perseus Books.

Dulac, N. 2007. *A Framework for Dynamic Safety and Risk Management Modelling in Complex Engineering Systems*, PhD Dissertation. Cambridge, MA: MIT.

Faraj, S., Kwon, D. and Watts, S. 2004. Contested artefact: Technology sensemaking, actor networks, and the shaping of the web browser. *Information Technology and People*, **17**(2), 186–209.

Flach, J.M., Dekker, S. and Stappers, P.J. 2008. Playing twenty questions with nature (the surprise version): Reflections on the dynamics of experience. *Theoretical Issues in Ergonomics Science*, **9**(2), 125–54.

Flood, R.L. 1999. *Rethinking the Fifth Discipline: Learning within the Unknowable*. London: Routledge Publishing.

Gephart, R.P., Van Maanen, J. And Oberlechner, T. 2009. Organizations and risk in late modernity. *Organization Studies*, **30**(2–3), 141–55.

Gherardi, S. and Nicolini, D. 2000. The organizational learning of safety in communities of practice. *Journal of Management Inquiry*, **9**(1), 7–18.

Grote, G. and Künzler, C. 2000. Diagnosis of safety culture in safety management audits. *Safety Science*, **34**, 131–50.

Gupta, S. 2010. Gulf oil spill report blames industry and regulators. *New Scientist*. 18 November 2010. Available at: http://www.newscientist.com/article/dn19739-gulf-oil-spill-report-blames-industry-and-regulators.html [accessed: 19 November 2010].

Hammer, D. 2010. Federal Investigators blast BP over 'safety culture' at oil spill hearings. *The Times-Picayune*. 26 August 2010. Available at: http://www.nola.com/news/gulf-oil-spill/index.ssf/2010/08/federal_investigators_blast_bp.html [accessed: 19 November 2010].

Harbers, H. 2005. *Inside the Politics of Technology: Agency and Normativity in the Co-production of Technology and Society*. Amsterdam: Amsterdam University Press.

Hernes, T. 2005. The Organization as a Nexus of Institutional Macro Actors: The Story of a Lopsided Recruitment Case, in *Actor-Network Theory and Organizing*, edited by B. Czarniawska and T. Hernes. Malmo: Elanders Berlings, 112–28.

Hughes, S. 2010. Spill Panel says rig culture failed on safety. *The Wall Street Journal*, 10 November 2010. Available at: http://online.wsj.com/article/SB10001424052748704635704575604622510434324.html [accessed: 19 November 2010].

Interim Report on Causes of the Deepwater Horizon Oil Rig Blowout and Ways to Prevent Such Events. National Academy of Engineering and National Research Council. Available at: http://www.nap.edu/catalog/13047.html [accessed: 25 November 2010].

Latour, B. 1987. *Science in Action: How to Follow Scientists and Engineers Through Society*. Milton Keynes: Open University Press.

Latour, B. 1992. 'Where are all the Missing Masses? The Sociology of a Few Mundane Artefacts', in *Shaping Technology/Building Society: Studies in Sociotechnical Change*, edited by W.E. Bijker and J. Law. Cambridge, MA: MIT Press, 225–58.

Latour, B. 1994. Pragmatogonies: A mythical account of how humans and nonhumans swap properties. *Behavioural Scientist*, **37**(6), 791–808.

Latour, B. 1996. On actor-network theory: A few clarifications. *Soziale Welt*, **47**, 369–81.

Latour, B. 1999. *Pandora's Hope: Essays on the Reality of Science Studies.* Cambridge, MA: Harvard University Press.

Latour, B. 2005. *Reassembling the Social: An Introduction to Actor Network Theory.* Oxford: Oxford University Press.

Lavelette, M. and Wright, C. 1991. The Cullen Report-making the North Sea safe? *Critical Social Policy,* **11**(31), 60–69.

Law, J. 1992. Notes on the theory of the actor network: Ordering, strategy and heterogeneity. *Systems Practice,* **5**(4), 379–93.

Law, J. 1999 After ANT: Complexity, Naming and Topology, in *Actor Network Theory and After,* edited by J. Law and J. Hassard. Oxford: Blackwell, 1–14.

Lockie, S. 2004. Collective agency, non-human causality and environmental social movements: A case study of the Australian 'landcare movement'. *Journal of Sociology,* **40**(1), 41–58.

Loermans, J. 2002. Synergizing the learning organization and knowledge management. *Journal of Knowledge Management,* **6**(3), 285–94.

Macalister, T. 2006. Shell accused over oil rig safety. *The Guardian* 23 June 2006. Available at: www.guardian.co.uk/business/2006/jun/23/oilandpetrol. freedomofinformation [accessed: 1 June 2010].

Masys, A.J. 2010. Fratricide in Air Operations: Opening the Black Box: Revealing the Social, PhD Dissertation. Leicester: University of Leicester, UK.

Mearns, K., Flin, R., Gordon, R., and Fleming, M. 1998. Measuring safety climate on offshore platforms. *Work and Stress,* **12**, 238–54.

Mearns, K., Whitaker, S.M., and Flin, R. 2003. Safety climate, safety management practice and safety performance in offshore environments. *Safety Science,* **41**, 641–80.

Miller, K. 1991. Piper Alpha and the Cullen Report. *Industrial Law Journal,* **20**(3), 176–87.

Miller, K.D. 2009. Organizational Risk after Modernism. *Organization Studies,* **30**(2–3), 157–80.

Monteiro, E. 2000. Actor Network Theory and Information Infrastructure, in *From Control to Drift: The Dynamics of Corporate Information Infrastructures,* edited by C. Ciborra. Oxford: Oxford University Press, 71–83. Available at: www.idi.ntnu.no/~ericm/ant.FINAL.htm [accessed: 11 July 2005].

Murdoch, J. 1997. Inhuman/nonhuman/human: Actor-network theory and the prospects for a nondualistic and symmetrical perspective on nature and society, *Environment and Planning D: Society and Space,* **15**(6), 731–56.

Newman, B. 2005. Agents, Artefacts and Transformations: The Foundation of Knowledge Flows, in *Handbook on Knowledge Management: Knowledge Matters,* edited by C.W. Holsapple. Heidelberg: Springer Publishing, 301–16.

Pate-Cornell, M.E. 1993. Learning from the Piper Alpha Accident: A Postmortem Analysis of Technical and Organizational Factors. *Risk Analysis*, **13**(2), 215–32.

Perrow, C. 1984. *Normal Accidents: Living with High-Risk Technologies*. New York, NY: Basic Books, Inc.

Perrow, C. 1999. *Normal Accidents: Living with High-Risk Technologies* (2nd edition). Princeton, NJ: Princeton University Press.

Power, S., Casselman, B. and Gold, R. 2010. Gulf spill linked to BP's lack of discipline. *The Wall Street Journal*, 17 November 2010.

Reason, J. 2004. Beyond the organisational accident: The need for 'error wisdom' on the frontline. *Quality and Safety in Healthcare*, **13**(2), 28–33.

Reynolds, C. 1987. Flocks, birds, and schools: A distributed behavioural model. *Computer Graphics*, **21**, 25–34.

Rosen, P. and Rappert, B. 1999. The culture of politics and technology. *Soziale Technik*, **4**(99), 19–22.

Senge, P. 1990. *The Fifth Discipline: The Art and Practice of the Learning Organization*. New York: Doubleday Currency.

Senge, P. 2006. *The Fifth Discipline: The Art and Practice of the Learning Organization*. New York: Doubleday Currency.

Snook, S.A. 2000. *Friendly Fire: The Accidental Shootdown of US Black Hawks Over Northern Iraq*. Princeton, NJ: Princeton University.

Somerville, I. 1997. Actor network theory: A useful paradigm for the analysis of the UK cable/on-line socio-technical ensemble? Available at: http://hsb/baylor.edu/eamsower/ais.ac.97/papers/somervil.html [accessed: 10 August 2004].

Stanton, N.A., Salmon, P.M., Walker, G.H. and Jenkins, D.P. 2009. Genotype and phenotype schemata and their role in distributed situation awareness in collaborative systems. *Theoretical Issues in Ergonomics Science*, **10**(1), 43–68.

Tatnall, A. 2000. *Innovation and Change in the Information Systems Curriculum of an Australian University: A Socio-technical perspective*, PhD Dissertation, Faculty of Education. Central Queensland University.

Toft, B. and Reynolds, S. 1999. *Learning from Disasters: A Management Approach* (2nd edition). Leicester: Perpetuity Press.

Turner, B.A. 1976. The organizational and interorganizational development of disasters. *Administrative Science Quarterly*, **21**(3), 378–97.

Turner, B.A. 1978. *Man-made Disasters*. London: Wykeham.

Turner, B.A. and Pidgeon, N.F. 1997. *Man-Made Disasters* (2nd edition). Oxford: Butterworth and Heinemann.

Usuki, M. and Sugiyama, K. 2003. Visualization methods for sharing knowledge pieces and relationships based on biological models, KES2003, Oxford, 3–5 September 2003, Springer LNAI 2774, 786–93.

Van der Duim, V.R. 2005. Tourismscapes, an actor-network perspective on sustainable tourism development, Dissertation Wageningen University, 29 June 2005.

Vaughan, D. 1996. *The Challenger Launch Decision: Risky Technology, Culture and Deviance at NASA*. Chicago, IL: University of Chicago Press.

Viseu, A.A.B. 2005. Augmented Bodies: The Visions and Realities of Wearable Computers, PhD Dissertation. University of Toronto.

Weick, K.E. and Sutcliffe, K.M. 2007. *Managing the Unexpected: Resilient Performance in an Age of Uncertainty* (2nd edition). San Francisco, CA: John Wiley and Sons Inc.

White, D. 1995. Application of systems thinking to risk management: A review of the literature. *Management Decision*, **33**(10), 35–45.

Wickramasinghe, N., Tumu, S., Bali, R.K. and Tatnall, A. 2007. Using Actor Network Theory (ANT) as an analytic tool in order to effect superior PACS implementation. *International Journal of Networking and Virtual Organizations*, **4**(3), 257–79.

William-Jones, B. and Graham, J.E. 2003. Actor-Network Theory: A tool to support ethical analysis of commercial genetic testing. *New genetics and Society*, **22**(3), 271–96.

Woods, D. 2003. *Creating Foresight: How Resilience Engineering Can Transform NASA's Approach to Risky Decision Making*. Testimony on the Future of NASA for Committee on Commerce, Science and Transportation, John McCain, Chair, 29 October 2003.

Yeung, H.W.C. 2002. Towards a relational economic geography: Old wine in new bottles? Paper presented at the 98th Annual meeting of the Association of American Geographers, Los Angeles, CA, 19–23 March 2003. Available at: http://courses.nus.edu.sg/course/geoywc/publication/Yeung_AAG.pdf [accessed: 1 September 2009].

4

The Inhuman: Risk and the Social Impact of Information and Communication Technologies

David Alford

Introduction

Here in the early stages of the twenty-first century, information and communication technologies (ICTs) are so deeply woven into the fabric of everyday activity that it requires a stretch of the imagination to conceive of life without them. Considering the apparently unstoppable growth of these technologies (for example, Moore 1965, Gilder 2000), one might reasonably ask: when do useful 'tools' become 'participants' – or even begin to enslave us? Some theorists think that the irresponsible overreaching of technology is dangerous and that its development and use need to be governed. In this debate it is pertinent to consider the evolution of ICTs and what characteristics might make them especially risky. If indeed there are substantial risk issues to be addressed, one might then ask how we approach regulation, ethics, and the possible role of public participation.

In *The Inhuman* Jean-Francois Lyotard (1991) presents a critique of technological development and its threat. In this context Inhumanism is a term to cover cases where the human dimension is eclipsed by the technological, or taken to be subsidiary to it in some way (Sim 2001). In Ulrich Beck's *Risk Society* (1992), the author describes a reflexive modernity. In reflexive modernity risks originate not just in nature, but paradoxically also in science and technology. We see a 'boomerang effect' where new risks rebound upon society at large. As Beck puts it, 'Society becomes a laboratory, but there is no-one responsible for its outcomes'. Beck suggests that the modernist project, funded by global capital, underwritten by scientific and technological innovation, and facilitated (enabled) by bureaucracies and academic and political elites, has become dysfunctional. Modernity seems more able to multiply than eliminate hazards. Lyotard, in *The Inhuman* (1991), observes that:

> The human race is 'pulled forward' [by the development of the techno-scientific system] without possessing the slightest capacity for mastering it. It is even probable that this has always been the case throughout human history. And if we can become aware of that fact today, this is because of the exponential growth affecting sciences and technology.

In *The Postmodern Condition* (Lyotard 1984) – a study of the nature of knowledge in computerized societies – he depicts the decline of the Grand Narrative, and expresses skepticism towards universal explanatory theories. Further, Lyotard sees contemporary science as no longer a coherent truth-orientated pursuit of knowledge, but rather as an array of 'language games' in which the search for desirable truths no longer counts, but only 'performativity' (instrumental functioning). Performativity suggests circularity – the research that will work best is that which generates more research and the acquisition of funding and power. Under the conditions of a computerized information society the great 'metanarratives' by which modernity had been legitimized are no longer sustainable. In addition ethical considerations are marginalized. As Lyotard (1991) remarks:

> Development is not attached to an Idea, like that of the emancipation of reason and of human freedoms. It is reproduced by accelerating and extending itself according to its internal dynamic alone.

Stuart Sim's book *Lyotard and the Inhuman* (2001) gives an account of Lyotard's misgivings about modern technology's agenda. Lyotard is deeply opposed to any shift towards the Inhuman – which has the backing of the forces of 'TechnoScience' (technology plus advanced capitalism and the multinationals). From Lyotard's viewpoint there is a threat that TechnoScience may impose its programme on us by stealth. There is a progressive blurring of boundaries between human and non-human, as Sim writes:

> The relationship between human and machine has altered dramatically in recent decades. Where once that relationship was one of domination with humans firmly in control, increasingly it has become one of cooperation – and even sometimes of domination from the machine side (particularly so when it comes to the more sophisticated forms of AI).

How far we are willing to allow the latter to continue developing is an
interesting moral dilemma – arguably the most important dilemma of
our age.

Efficiency and enhanced performance are what drive development. Lyotard
is unsettled by the obsessive goal-directedness and efficiency of modern
technology, spurred by time-saving and profit motives. Gane (2003) remarks
that there is, for Lyotard, an intimate connection between digital technologies
and relations of power. These technologies may be regarded as extensions of
the capitalist market in that they accelerate and so promote the production,
exchange and consumption of information. To what extent is this undesirable
or uncontainable, and does it lead to the erosion of morality? Perhaps – Lyotard
suggests – the most important trait that inhumanism seeks to eradicate is
that of 'difference' or heterogeneity: the human mind is anarchic and resists
containment, and we must treasure the flexibility of human thought and
our analogical reasoning faculties, compared with the rigid, inflexible logic
of computer systems (Sim 2001). As Gane (2003) elucidates – technological
structures impose rationalization on thought. This process is part of a general
homogenization of all cultural forms. As well as increasing the pace of life
under a regime of efficiency, performance and control, Gane (2003) explains
that the digital transformation of culture has a further consequence:

> *... in our day-to-day processing of short 'bytes' of information we*
> *ourselves become more like machines. In other words, through our*
> *use of new media technologies, we, as humans, become increasingly*
> *'inhuman'... [Lyotard's] position may be read as an extension of Marx's*
> *theory of the alienation of capitalist subjects from their underlying*
> *human nature ...*

RELIANCE ON COMPUTER SYSTEMS

The Millennium Bug (Y2K) demonstrated how much of our control had been
relinquished to computer systems; humans did not seem to be in charge.
Y2K portended a Doomsday scenario, which turned out to be a non-event.
Nevertheless, it highlighted our increasing reliance on computer systems –
pervading every aspect of modern living. The pace and extent of technological
change can lead to both euphoria and 'technophobia' (see, for example,
Sandywell (2006). As described in Gere's (2008) book *Digital Culture*, Y2K
indicated alarming potential consequences: the breakdown of – for example –
banking, welfare distribution, medical equipment, electricity grids, air traffic

control and almost any system that uses digital technology (any developed nation's infrastructure), perhaps even the accidental launch of nuclear missiles. In the end – as we know – nothing happened. Nevertheless the issues that emerged concerning our increasing reliance on complex systems remain a definite cause for concern (see also Kuflik (1999) for a moral perspective on ceding control to computers).

So how did we get where we are? Gere (2008) describes the beginning of digital culture:

> *The Second World War was the catalyst not just for the invention of the modern binary digital electronic computer, but also for the development of a number of remarkable and influential discourses, including Cybernetics, Information Theory, General Systems Theory, Molecular Biology, Artificial Intelligence, and Structuralism.*

These are abstract, formalized systems – the paradigm of post-war technological and scientific thinking. Reflecting on the cyclical nature of capitalism, Mandel (1975) traces the latest technological revolution back to 1940 – a revolution defined by the control of machines by electronic apparatus (and also the general introduction of nuclear energy). Tavani (2001, 2007) gives an account of four phases of computing technology – evolving from stand-alone machines to complete technological convergence:

- Phase 1 (1950s and 1960s): the era of unconnected mainframe computers. Potential social concerns arising were electronic data gathering (the so-called 'Big Brother' society), and the consequences of developing artificial intelligence.

- Phase 2 (1970s and 1980s): the beginnings of computer networks. Potential social concerns arising were personal privacy, intellectual property and computer crime.

- Phase 3 (1990 to the present): the internet era. Potential social concerns centre around free speech, anonymity and trust.

- At present we are on the threshold of what could be described as Phase 4: a high level of technological convergence, and ubiquitous or 'pervasive' devices.

The consequence of this latter phase are not really well understood, but it is a widely held belief that technological convergence and ubiquity may lead to aspects of our biology and our technology merging or becoming less distinct (see, for example, Moor (2005a)). Tavani (2007) elaborates on the possible implications of Phase 4:

> *Many technologies that were previously distinct will most likely converge, and the pace at which technological convergence occurs will most likely become accelerated [...] As convergence in the fields of biotechnology and IT continues, it is not clear whether computers of the future will be silicon-based or whether they will possibly be made also of biological materials [...] many predict that computers will become increasingly smaller in size, ultimately achieving the nanoscale. Many also predict that nanotechnology, biotechnology and IT will continue to converge.*

The development of artificial intelligence (AI) – at least as the concept was originally understood – has so far failed to live up to early expectations. But, as Gere (2008) observes, more recent developments, many of which came out of AI, are presenting us with objects and technologies that can act, communicate, signify and participate, even if these capacities do not seem anything like human intelligence or consciousness. It seems unlikely in the foreseeable future that there will exist even minimally intelligent robots. But smaller developments are already provoking questions about the capacity for technology to act and participate. RFID (radio frequency identification) is a new technology that allows data to be automatically sent from objects that contain RFID tags, used for example in tagging library books, public transport travel passes, passports, traffic toll systems, stock inventory, prisoner control and many other uses. The use of RFID chips is controversial (indeed the devices have become known as 'spychips'). Digital technologies are no longer just tools, but increasingly are participants, for better or worse – and the technology is becoming increasingly invisible as it becomes more a part of our existence (described as 'pervasive' or 'ubiquitous' technologies). As for AI, even ardent AI devotees can foresee potential problems with the creation of super-AI entities called 'artilects', by 'evolutionary engineering': 'The issues of massive intelligence will dominate global politics of the [21st] century' (Hugo de Garis, 1997 – cited in Sim, 2001). Noel Sharkey (2009, see also Fleming and Sharkey 2009 used to be a believer in AI – now he thinks AI is a dangerous myth that could lead to a dystopian future. Sharkey believes that 'there is no evidence of an artificial toe-hold in sentience [...] the mind could be a system that cannot be recreated by a computer' (but see contrary viewpoints from Hans Moravec

(1998) and Ray Kurzweil (1999)). However, Sharkey is concerned about the development of military robots – the irresponsible over-reaching of technology may be dangerous, and if so we need ethical guidelines to govern the use of robots.

Lyotard (1991) concedes that, despite the primitive nature of AI as it stands, progress can be expected in information science, artificial languages and communications science. Computers are starting to be able to deliver simulacra of certain mental operations. But his main objection concerns the very principle of these intelligences. Citing Hubert L. Dreyfus (1965, 1972) he notes:

> ... our disappointment in these organs of bodiless thought comes from the fact that they operate on binary logic, one imposed on us by Russell and Whitehead's mathematical logic, Turing's Machine, McCulloch and Pitt's neuronal model, Boolean algebra, and Shannon's information science. But as Dreyfus argues, human thought doesn't think in a binary mode. It doesn't work with units of information (bits), but with intuitive hypothetical configurations [...] Human thought can distinguish the important from the unimportant without doing exhaustive inventories of data and without testing the importance of data with respect to the goal pursued by a series of trials and errors.

Perri (2001) gave an account of the manner in which fears about AI cluster around several themes, of which three are of central importance:

- In the extreme variant the fear is of AI as all-powerful, or as the tyrant out of control.

- Secondly, the opposite fear of AI as highly vulnerable is also evident. This is the fear that we shall find these systems so benign and so reliable that we shall become so routinely dependent upon their operations, with the result that any systemic failure – whether resulting from natural causes, from technical weaknesses, from the incomprehensibility of inherited systems making repair almost impossible, or from malevolent human guided attack by information warfare – could bring our society to its knees.

- Thirdly, there are Weizenbaum's (1984) fears that we shall allocate categories of decisions to artificially intelligent systems, resulting both in the degradation of values and in damaging outcomes. As a result of using AI to make decisions, both human capabilities of, and the social sense of moral responsibility for, making judgement will be corroded.

Moreover, in addressing the deep-seated ramifications of extensive and widespread use of ICTs, there is the matter of how computers may alter the nature of human cognition. ICTs may alter the character of human thought processes, interaction, risk perception, judgement and decision-making and promote unnatural haste. Gane (2003) expresses this perfectly:

> The digitalization of data tears both cultural artifacts and sensory experience from their moorings in physical time and space. The result is what Lyotard terms a hegemonic teleculture.

'Virtuality' and telecommunications technology change our perceptions, annihilate physical distance and dissolve material reality. Baudrillard (1994) presents a vision of the world where meaning is destroyed by the act of communication itself – the death of the 'Real'. Here, the distinction between representation and reality – (i.e. between signs and what they refer to in the real world) – breaks down. Reality becomes redundant and we have reached 'hyper-reality', in which images breed incestuously with each other without reference to reality or meaning.

A threat that is increasingly surfacing in press reports and news bulletins is that of cyberterrorism. As described in Tavani (2007), examples of imagined cyberterror scenarios would be politically motivated hacking operations designed to cause grave harm such as bringing down financial markets, or taking control of public transportation systems to bring about collisions. Perhaps what is most disquieting about these potential threats is that – owing to the complexity of modern computer networks – it can be very difficult to distinguish network disruption due to malicious hacking ('cybervandalism' or pranks) from cyberterrorism, or even whether the disruption has come about merely through system failure. The topic may be considered in tandem with the topic of information warfare (see, for example, Denning (1999)).

ORGANIZATIONAL RELIABILITY

When considering the potential social and organizational impact of ICTs, a couple of publications spring to mind concerning:

1.   the concept of 'slack'; and

2.   'practical drift' in relation to software vulnerability.

Schulman (1993) discusses the usefulness of slack in organizations with reference to a case study on the Diablo Canyon Nuclear power plant:

> 'Slack' [...] implies looseness in the organization of things – one part of a system trailing badly the activity of another. Yet slack is a critical, if underappreciated, managerial resource. It is slack that provides a margin of error between a lapse in one aspect of an organisation's performance and harmful consequences in every other. It is slack that allows managers the freedom to manoeuvre – to act decisively on one part of a problem without having their decisions ramify quickly, widely and unexpectedly to every other part.

With resource slack (to do with time, money, personnel) the surplus of these values withheld from commitment can be seen as non-productive suboptimal inefficacy – but can also be seen as a hedge against the unexpected. Control Slack (individual degrees of freedom in organizational activity) can be viewed as a lack of 'crispness', but can also be viewed as protection against the dysfunctions of centralised authority. Schulman notes that modern communications and computer networks generate pressure for rapid-fire organizational responses and time-critical decisions. Reliability in particular has become an important requirement for organizations. Managers attempt to 'lock in' practices they can count on to hold their organizations within the bounds of 'safe' operation – in pursuit of reliability managers instinctively push towards strategies that threaten even the last vestiges of organizational slack by developing control systems that seek to promote invariant behaviour.

It is accepted by everyone that nuclear plants are a complex and inherently hazardous technology. But what is not widely appreciated is just how daunting an administrative and organizational task it is to manage them safely. The third kind of slack is what Schulman terms 'conceptual slack' – a divergence in analytical perspectives among members of an organization over theories, models or casual assumptions pertaining to its technology or production processes. This may be viewed as confusion or ambiguity, but it can also be seen as a protection against 'errors of rendition' or an aversion to errors of 'aggressive hubris'. Schulman concludes that the toleration, even the protection, of organizational ambiguities allows an organization to cope with the unforeseen. Managers work against this objective when they associate reliability with invariance.

Lundestad and Hommels (2007) discuss the vulnerability of software due to 'practical drift' – a theoretical perspective developed by Snook (2000) in a military context. When a system becomes used, it acquires a degree of flexibility. Then 'drift' occurs – safety instructions, built into the system by its designers, gradually recede to the background. As a result the system becomes more and more vulnerable ... 'the slow, steady uncoupling of practice from written procedure' (Snook 2000). Lundestad and Hommels (2007) suggest that the way forward lies in accepting the risks associated with our reliance on increasingly complex computer systems and in shedding our naive faith in scientific and technological solutions that are supposed to make complex systems risk-free. Software developers could also be trained to recognize the social nature of their work, rather than seeing it as a purely technical task.

PUBLIC PARTICIPATION AND THE DEMOCRATIZATION OF KNOWLEDGE

Modern approaches to risk, crisis and disaster management give pre-eminence to expert knowledge and the formal state apparatus of disaster response and recovery. Late-modern approaches, in contrast, recognise lay knowledge and the benefits of public participation. Many scientists and experts on the public understanding of science argue for the improvement of citizens' scientific literacy so that decision-making can be based on the best available knowledge (see Prewitt (1983) and Miller (cited in Heise 2004)). Irwin (1995) suggests that employees' knowledge and insights can help improve technology design assessment and implementation. He also suggests that contemporary publics want more influence over which technologies are developed and how they are used, a concept he terms 'technological citizenship'. Zimmerman (1995) discusses a 'democratic ethic of technological governance'. This eloquent discourse is largely in the context of nuclear power as a 'prototypical example of technological authoritarianism [sponsored by] ... the convergence of national security and entrenched economic interest'. However, the arguments hold universally for large-scale technological ventures. Zimmerman poses important questions of how autonomous development comes about among the adult members of society. Are we concerned about (or even aware of) technology curtailing our autonomy? If political forms are embedded in various technologies – how might this relate to the development of the individual? Can we structure both technological design and our own moral development in line with democratic principles? Addressing the problem of technological authoritarianism Zimmerman cites Langdon Winner:

> *Consciously or unconsciously, deliberately or inadvertently, societies choose structures for technologies that influence how people are going to work, communicate, travel, consume, and so forth over a very long time …. Because choices tend to become social habit, the original flexibility vanishes for all practical purposes once the initial commitments are made. In that sense technological innovations are similar to legislative acts or political foundings that establish a framework for public order that will endure over many generations. (Winner 1986)*

Artifacts have politics, and the well-being and freedom of those obliged to participate in technology is a function of the technological system itself (Zimmerman 1995).

Frankenfeld (1992 cited in Zimmerman (1995)) defined technological citizenship as 'equal membership, participation, and standing or status of persons as agents and subjects within a realm of common impact to at least one "technology" or instance of consciously amplified human capacity under a definable state that governs this technology and its impacts. Such status is defined by a set of binding, equal rights and obligations that are intended to reconcile technology's unlimited potentials for human benefit and ennoblement with its unlimited potentials for human injury, tyrannization, and degradation'. Accordingly 'the overarching goals of citizenship are (1) autonomy, (2) dignity, and (3) assimilation – versus alienation – of members of the polity'.

There are, of course, issues with the late-modern approach. For example, do members of the public have the requisite knowledge to be equal participants in complex debates over risk assessment and management? In this regard Norman Levitt considers genuine scientific literacy to be an unattainable goal for the majority of people; he believes their knowledge will remain limited. From this stance the public would still have a role in deciding what kinds of broad purposes science should serve and what kind of projects should be prioritized. But the public's task is not to acquire expert knowledge, but rather to learn how to discriminate between experts and non-experts, and how to seek the more useful opinions of the former (Levitt cited in Heise 2004). In addition, one might reasonably ask whether busy working people have the time to attend meetings, hearings, inquiries and participate in site visits? Finally there may be issues of commercial confidentiality or national security to be considered.

At this point it may be useful to consider two examples demonstrating how the public have been involved in ICT assessment:

    i.   in the field of security and surveillance; and

    ii.   in the field of nanotechnology and human enhancement.

(I) THE EUROPEAN TECHNOLOGY ASSESSMENT PROJECT (PRISE) – JOHANN ČAS (2010) IN ASSOCIATION WITH THE INSTITUTE OF TECHNOLOGY ASSESSMENT, VIENNA

This study investigated attitudes towards different groups of technologies. The technologies are biometrics, camera surveillance (CCTV), scanning technologies, locating technologies, eavesdropping, data retention and privacy-enhancing technologies. The project carried out six so-called 'interview meetings' in Austria, Denmark, Germany, Hungary, Norway and Spain. The interview meeting is a method that combines debate, completing a questionnaire and group discussions. An interview meeting takes three hours and is normally held as an after-work event. The six interview meetings of the PRISE project resulted in six national reports (summarized in Čas (2010)). There were minor variations across the different nationalities with respect to attitudes to security technologies and trust in governments. Notwithstanding these national differences majority consensus opinions can be summarized thus: the threat of terror as such does not justify privacy infringements; physically intimate technologies are unacceptable; misuse of technology must be prevented; function creep is not acceptable (this refers to technologies or data being used for something other than the original purpose). The ensuing democratic demands were as follows:

> *Public debate* – decisions on implementing new security technologies or measures must always be based on a transparent decision-making process. More importantly – before these kinds of decisions are taken, there must always be an informative and involving debate.

> *Broad involvement* – all relevant parties, including experts and human rights organizations, must be heard prior to important decisions on security and privacy.

> *Always analyse privacy impact* – before implementing new security technologies the privacy impact of the technologies must be analysed

thoroughly. Funding of research projects on new security technology should also be dependent on an analysis of the possible privacy impact.

(II) NATIONAL CITIZENS' TECHNOLOGY FORUM (NCTF) ON NANOTECHNOLOGY AND HUMAN ENHANCEMENT (2008) – IN SCLOVE (2010)

The forum is in association with the Expert and Citizen Assessment of Science and Technology (ECAST) network. ECAST is proposed as being independent of the government and comprising a complementary set of non-partisan policy research institutions, universities and science museums across the United States. The NCTF is important as the first US participatory technology assessment (pTA) exercise to be organized on a national scale.

The study examined the implications of radically enhancing individual human capabilities through the combined use of biotechnology, nanotechnology, information technology and cognitive science. The NCTF took place in March 2008, and included 74 lay participants from six states (New Hampshire, Georgia, Wisconsin, Colorado, Arizona and California). All participants received a 61-page background document, and then met face-to-face at their respective sites for a full weekend at the beginning of the month and for a concluding weekend at the end of the month. In between, the lay panelists participated together in nine two-hour internet sessions, posing questions online to a panel of experts, sharing concerns and perspectives among the sites and selecting and refining the common set of questions that would guide the culminating face-to-face session.

As a result of participating in the NCTF, the proportion of lay panelists anticipating that the social benefits of enhancing human capabilities would exceed the risks declined from 82 per cent to 66 per cent. All six sites registered significant concern about the effectiveness of government regulations for human-enhancement technologies and recommended better public information, education and deliberation about these technologies. Five of the six sites assigned higher priority to funding treatment of diseases than to enhancement research, and they also advocated stakeholder involvement in setting research agendas. The organizers later described the process and findings of the NCTF at a congressional briefing, and there are indications that the NCTF may have influenced language mandating 'deliberative public input in decision-making processes' within a 2009 US Senate bill seeking to reauthorize the National Nanotechnology Initiative.

ETHICS AND PRECAUTIONS

The general field of technoethics studies the moral and ethical aspects of technology in society (Galván 2003, Bao and Xiang 2006, Luppicini and Adell 2008). In the context of ICTs some authors employ the inclusive term 'cyberethics' when referring to interconnected communications technologies (Tavani 2007). In James Moor's (2001, 2005b) papers on the future of computer ethics he makes the point that computers are unique, for they are, at least in principle, universal machines like no other:

> Computers are logically malleable machines in that they can be shaped to do any task that one can design, train or evolve them to do. Syntactically, computers are logically malleable in terms of the number and variety of logical states and operations. Semantically, computers are logically malleable in that the states and operations of a computer can be taken to represent anything we wish.

Applications for computers seem to be abundant and ever-increasing. As a result of the widespread applications of computers, novel opportunities are created. In many cases the uses of computers are so unusual that no policies for their proper use exist – or even have been considered. Moor proposes the need to formulate policies, even if only informally, to make sure our actions remain within ethical bounds:

> ... in general, the larger the number of novel applications of computer technology, the larger the number of problems in computer ethics ... as computing capacities increase exponentially, computing applications with associated policy vacuums will as well.

In 'Why we need better ethics for emerging technologies', Moor (2005b) brings into focus three other rapidly developing technologies: genetic technology, nanotechnology and neurotechnology. These are three rapidly developing technological movements. None of the three has progressed as far as computer technology in terms of its impact on society, but each has tremendous transformative – even revolutionary – potential, and in addition these technologies tend to be convergent. As Moor elaborates, these technologies could, at some point in the future, offer us the means to build new bodies, new environments, and even new minds. Such research activities are very likely to be funded, and to be developed. Moor points out that our ethical understanding of developing technology will never be complete, but can do much to 'unpack the potential consequences of new technology', while realising that ethics is an ongoing and dynamic enterprise. As sometimes suggested, another possibility is to place a moratorium on technological development until ethics catches up. Joy (2000) takes this stance, remarking that:

> The 21st-century technologies – genetics, nanotechnology, and robotics (GNR) – are so powerful that they can spawn whole new classes of accidents and abuses. Most dangerously, for the first time, these accidents and abuses are widely within the reach of individuals or small groups. They will not require large facilities or rare raw materials. Knowledge alone will enable the use of them.

The second improvement that would make ethics better would be establishing better collaborations among ethicists, scientists, social scientists and technologists. We need a multi-disciplinary approach (Brey 2000).

The area of nanotechnology and nanocomputing is of particular interest from an ethical viewpoint. These technologies, at the time of writing, are largely hypothetical; indeed there is no universally agreed definition of what nanotechnology constitutes. However, accepting that the field is at present just a kind of 'thought exercise', the ramifications of such potential technological developments are fascinating. In the 1980s Eric Drexler predicted that developments in nanotechnology would result in computers at the nano-scale (Drexler 1986, 1991). Drexler also proposed the idea of nano-scale 'assemblers', programmed to be self-replicating. There have also been notions of quantum nanocomputers, and the concept has even been floated of using DNA strands as building blocks for machines at this scale (see Tavani 2007).

As already mentioned, the whole field is a matter of speculation – suffice it to say that merely using conventional computing technologies, standard computing chips have already been constructed at the nano-scale.

Som, Hilty and Ruddy (2004) raise the important and interesting topic of the Precautionary Principle (PP), and how this might apply to ICTs. As new technologies emerge – one must evaluate opportunities in tandem with risk. Som et al. regard the PP as providing a framework to anticipate and minimize the risks of novel technologies as well as to foster their positive potential. Before outlining their approach to ICTs here is a brief background on the PP: initially it grew out of the environmental awareness movement in the late 1960s. Löfstedt, Fischoff and Fischoff (2002) report the first legal use of the general concept in the 1969 Swedish Environmental Protection Act. It was significant in reversing the burden of proof – requiring industry to demonstrate the safety of environmentally hazardous activities, rather than the onus being on critics to demonstrate negative consequences. The German government developed a weaker version of the PP, *Vorsorgungsprinzip* (literally meaning 'showing prior care or worry'), implying good husbandry (Boehmer-Christiansen 1994). Van den Belt (2003) says that 'humanity finds itself in a historically unprecedented situation in which our technological capacity and the potential scale of our actions far exceed our predictive knowledge'. Van den Belt references Jonas's (1984) argument that we should 'give in matters of a certain magnitude – those with apocalyptic potential – greater weight to the prognosis of doom than to that of bliss'.

In essence the PP can be summarized by the aphorism 'better safe than sorry'. There are several versions and strengths of the PP, but the basis of it is the ethical principle that in cases where the consequences of an action – especially the use of new technologies – *may* be judged unsafe, then one should err on the side of not carrying out the action, rather than risking the negative consequences. It must be noted that the PP is somewhat controversial and not without its detractors who regard it as stifling, meddlesome or impractical (for example Sunstein (2002), Whyte (2007) and Clarke (2009); for a defence of the PP see Weckert (2010)). Som *et al.* describe the PP as a rather *open* principle. The 'openness' of the wording leaves room for democratic decisions, but also for arbitrary decisions and makes it difficult to operationalize the PP. Therefore there is an intensive debate going on about the utility of the PP (see, for example, Santillo and Johnston (1999), Graham (2000) and Tickner *et al.* (2003)). At the outset the PP was applied to environmental concerns. Increasingly the scope of the PP is being extended to include a variety of new technologies, such as artificial intelligence, robotics and nanotechnology (Weckert 2010).

THE CHARACTERISTICS OF ICTS

Som *et al.* (2004) summarize important characteristics of ICTs in relation to the PP:

1.   ICT is a mass consumer technology that enables the user to gain access to communication channels, information, and the control of processes.

2. ICT components will increasingly be embedded in other objects that will then take on 'intelligent' features, but will not be perceived as ICT devices.

3. There is a trend to interconnect the physical world (world of things) with the virtual world (world of data) in real time, that is, more and more data will be synchronized with physical processes via sensors, and *vice versa* via actuators. The opportunities this synchronization brings about for the organization of production and consumption processes tend to make us dependent on its availability and proper functioning.

4. The diffusion of novel ICT depends strongly on compatibility issues. The requirement that new ICT products remain compatible with existing ones narrows the range of future development trajectories. Therefore, the market development of ICT is highly path dependent. In fields where no open technical standards have been established (e.g., word processing), ICT markets tend toward a 'winner takes it all' structure.

5. ICT systems form complex distributed systems when networked. Responsibility for damages caused by ICT is as distributed as the technology itself. This leads to a dissipation of responsibility – a new version of the same 'organized lack of responsibility' that Ulrich Beck attributed to modern technologies. (Beck 1992)

Som *et al.* make the case that these characteristics mean that ICT would be expected to interact intensively with social practices, which may result in profound changes to social rules and structures in the near future. Already there are issues foreshadowing this development. For example the 'digital divide', where sections of society are excluded from ICT access, or the fact that privacy regulations are increasingly difficult to implement, because technology is advancing faster than the legal system can react.

Som *et al.* suggest that a basic precautionary heuristic is to preserve diversity and avoid path-dependency:

> *For the application of the PP to ICT, we can first conclude that open standards for all types of interfaces among ICT products are preferable over proprietary standards, because they are essential for avoiding strong*

*path-dependency and trends toward market dominance, which destroy
diversity. Secondly, less complex technical solutions should generally be
preferred to more complex ones, because unmastered technical complexity
fosters investment in analysis and adaptation, which fosters the path-
dependency of the development. Participation in a dialogue on novel
technologies, however, is only possible if there is a basic ability to reflect
on present technology and technological visions, which is usually lacking
in the ICT field.*

The authors thus conclude that a vital aspect of precaution lies in education;
there must be provision for a type of education that will give people the capacity
to critically reflect on ICT and its impacts. Jenkins (1999) describes how studies
on lay understanding of scientific concepts (for example Irwin and Wynne 1996)
have implications for the form and content of school science curricula. This
could, for example, mean helping students to engage reflexively with scientific
issues, considering areas where science is controversial, learning about how
scientific investigations are conducted, and promoting understanding of risk
assessment. As Jenkins (1999) remarks:

> *... it is perhaps also important to ask whether school science education
> can any longer encourage the view that the world is much simpler than
> it really is and, thereby, promote unsustainable claims about the power
> of science to explain and control.*

Conclusion

At the outset, this discussion focussed on the concept of the inhuman, and
Lyotard's suspicion of it. However, William Martin (2009) comments:

> *[W]hile it would be a mistake to describe the tone of The Inhuman as
> 'optimistic' (for the reason that Lyotard maintains his critique of the
> functional integration of science, technology and education in advanced
> capitalist societies), there is nevertheless an acceptance that the future
> of humanity will be determined by its capacity to negotiate a more
> creative, symbiotic relationship with these technologies.*

Martin remarks on the effects of workers becoming socialized to the
new technologies:

> *We can theorize the relationship between research and education as a dialect of dehumanization and rehumanization, for as Lyotard puts it, 'the system seems to be a vanguard machine dragging humanity after it, dehumanizing it in order to rehumanize it at a different level of normative capacity' [...]. What emerges from Lyotard's account of the postmodern condition is therefore the idea that the development of science and technology is governed by the internal logic of scientific discourse and the external logic of the market. Without the narrative of emancipation to counter the dehumanizing effects of technology, Lyotard has no option but to accept the advance of techno-science as the condition of both knowledge-production and socialization in 'computerized societies'.*

In opposition to Lyotard's theoretical position, Jürgen Habermas has reacted strongly to the postmodern notion of the end of modernity, proposing instead that we think of modernity as an unfinished project (Habermas 1985). He retains continued attachment to the ideal of rational enlightenment. He concedes that modernity and the techno-scientific enterprise has not lived up to its promise, but we need to keep hold of some of modernity's bold aspirations – to complete the modernist project, rather than abandon it.

Douglas Kellner (2005) notes that discourses on new technologies tend to be polarized as either technophilic (technology as salvation) or technophobic (technology as damnation). To some extent this has always been the case, for example with the introductory phases of telegraphy, film, radio and television. Kellner advocates a need for a critical theory of technology that steers a path between utopic and dystopic visions, providing a normative perspective that takes account of positive and negative uses as well as ambiguities.

In his discussion on the ethics of information technology Nuyen (2004) notes there is a prevailing view of postmodernism (typified by Lyotard's stance) as being an obstacle to a consensus of ethical principles. By dwelling on the fragmentation of societies, the denouncing of the Enlightenment grand narrative and rejection of unitary thought systems, some authors view postmodernism as a hindrance (Hamelink 2000, Garnham 2000). Conversely, Nuyen points out that a deeper reading of Lyotard's overall work is possible. In some situations certain language games assume positions of dominance; hence Lyotard asserts the moral imperative of maximizing the multiplication of small narratives (Lyotard and Thébaud 1985). From this approach stems freedom, invention and creativity. The more games there are to play, the less likely it is that any one game will totalize the field ... and so proceeds the 'setting up of the conditions of possibility for ethical discourse' (Nuyen 2004).

Public participation in the democratization of knowledge works as an ideal, but may experience practical implementation difficulties, especially in terms of members of the public achieving and maintaining the requisite knowledge to participate. Zimmerman (1995), being strongly opposed to technocratic imperialism, states that:

> *The eventual collapse of authoritarian technologies will only occur if the*
> *collective consciousness of society becomes more capable of recognizing*
> *this authoritarian tendency as an abdication of moral responsibility.*

He speaks of 'dismantling the more glaringly authoritarian technologies' in the eventual hope of restructuring other technologies. Zimmerman regards the democratic *process* of participation as crucial –however imperfect the resulting artifacts prove to be in the short term. This, in his view, will lead to longer-term improvement in technological structures 'as people's participation and moral development expand […] we must be careful not to become habituated to the initial choices because once technologies become well established, they tend to shape their surroundings and exact progressively greater degrees of conformity' (Zimmerman 1995).

The ethics of advanced technology is highly complex, due largely to the pace of development in the field and the potential convergence of multiple emerging technologies. One rather promising line of practical work would appear to be the application of the Precautionary Principle (Som *et al.* 2004), with regard to their classification of ICT characteristics.

References

Bao, Z. and Xiang, K. 2006. Digitalization and Global Ethics. *Ethics and Information Technology*, 8 (1), 41–7.

Baudrillard, J. 1994. *Simulacra and Simulation*. Michigan: University of Michigan Press.

Beck, U. 1992. *Risk Society: Towards a New Modernity*. London: Sage.

Boehmer-Christiansen, S. 1994. The Precautionary Principle in Germany – Enabling Government, in *Interpreting the Precautionary Principle*, edited by T. O'Riordan and J. Cameron. London: Earthscan, 31–60.

Brey, P. 2000. Method in Computer Ethics: Towards a Multi-Level Interdisciplinary Approach. *Ethics and Information Technology*, 2 (2), 125–9.

Čas, J. 2010. Privacy and Security: A Brief Synopsis of the Results of the European TA-Project PRISE, in *Data Protection in a Profiled World*, edited by S. Gutwirth. Dordrecht: Springer, 257–62.

Clarke, S. 2009. New Technologies, Common Sense and the Paradoxical Precautionary Principle, in *Evaluating New Technologies: Methodological Problems for the Assessment of Technological Developments*, edited by M. Duwell and P. Sollie. Dordrecht: Springer, 159–73.

D 5.8 Synthesis Report (2008). Interview Meetings on Security Technology and Privacy. [Online: PRISE Project – Privacy enhancing shaping of security research and technology – A participatory approach to develop acceptable and accepted principles for European Security Industries and Policies]. Available at: prise.oeaw.ac.at/docs/PRISE_D_5.8_Synthesis_report.pdf [accessed: 5 December 2010].

Denning, D.E. 1999. *Information Warfare and Security*. New York: ACM Press.

Dreyfus, H. 1965. *Alchemy and Artificial Intelligence*. [Online: RAND Corporation Paper]. Available at: http://www.rand.org/pubs/papers/2006/P3244.pdf [accessed: 5 December 2010].

Dreyfus, H. 1972. *What Computers Can't Do: A Critique of Artificial Reason*. New York: Harper and Row.

Drexler, E.K. 1986. *Engines of Creation: The Coming Era of Nanotechnology*. New York: Anchor/Doubleday.

Drexler, E.K. 1991. *Unbounding the Future: The Nanotechnology Revolution*. New York: Quill.

Fleming, N. and Sharkey, N. 2009. The Revolution Will Not Be Roboticised. *New Scientist*, 203 (2723), 28–9.

Frankenfeld, P. J. 1992. Technological Citizenship: A Normative Framework for Risk Studies. *Science, Technology and Human Values*, 17, 459–84.

Galván, J. 2003. On Technoethics. *IEEE Robotics and Automation Magazine*, 10 (4), 58–63.

Gane, N. 2003. Computerized Capitalism: The Media Theory of Jean-François Lyotard. *Information, Communication and Society*, 6 (3), 430–50.

Garnham, N. 2000. The Role of the Public Sphere in the Information Society, in *Regulating the Global Information Society*, edited by C.T. Marsden. New York: Routledge, 43–56.

Gere, C. 2008. *Digital Culture*. London: Reaktion Books.

Gilder, G. 2000. *Telecosm: How Infinite Bandwidth Will Revolutionize Our World*. New York: The Free Press.

Graham J. 2000. Perspectives on the Precautionary Principle. *Human and Ecological Risk Assessment*, 6, 383–5.

Habermas, J. 1985. *The Philosophical Discourse of Modernity*. Cambridge: Polity Press.

Hamelink, C. 2000. *The Ethics of Cyberspace*. London: Sage.

Heise, U. 2004. Science, Technology, and Postmodernism, in *The Cambridge Companion to Postmodernism*, edited by S. Connor. Cambridge: Cambridge University Press, 136–67.

Irwin, A. 1995. *Citizen Science*. London: Routledge.

Irwin, A. and Wynne, B. 1996. *Misunderstanding Science? The Public Reconstruction of Science and Technology*. Cambridge: Cambridge University Press.

Jenkins, E. 1999. School Science, Citizenship and the Public Understanding of Science. *International Journal of Science Education*, 21 (7), 703–10.

Jonas, H. 1984. *The Imperative of Responsibility: In Search of an Ethics for the Technological Age*. Chicago: University of Chicago Press.

Joy, B. 2000. Why the Future Doesn't Need Us. *Wired* 8.04 [Online]. Available at: http://www.wired.com/wired/wrchive/8.04/joy.html [accessed: 5 December 2010].

Kellner, D. 2005. New Technologies and Alienation: Some Critical Reflections, in *The Evolution of Alienation: Trauma, Promise, and the Millennium*, edited by L. Langman and D. Kalekin-Fishman. Maryland: Rowman and Littlefield, 47–68.

Kuflik, A. 1999. Computers in Control: Rational Transfer of Authority or Irresponsible Abdication of Autonomy? *Ethics and Information Technology*, 1, 173–84.

Kurzweil, R. 1999. *The Age of Spiritual Machines: How We Will Live, Work and Think in the New Age of Intelligent Machines*. London: Orion Business Books.

Levitt, N. 1999. *Prometheus Bedeviled: Science and the Contradictions of Contemporary Culture*. New Brunswick: Rutgers University Press.

Löfstedt, R., Fischoff, B. and Fischoff, I. 2002. Precautionary Principles: General Definitions and Specific Applications to Genetically Modified Organisms. *Journal of Policy Analysis and Management*, 21 (3), 381–407.

Lundestad, C.V. and Hommels, A. 2007. Software Vulnerability due to Practical Drift. *Ethics and Information Technology*, 9, 89–100.

Luppicini, R. and Adell, R. 2008. *Handbook of Research on Technoethics*. Hershey: Idea Group Publishing.

Lyotard, J-F. 1984. *The Postmodern Condition: A Report on Knowledge*. Manchester: Manchester University Press.

Lyotard, J-F. 1991. *The Inhuman: Reflections on Time*. Cambridge: Polity Press.

Lyotard, J-F. and Thébaud, J-L. 1985. *Just Gaming*. Minneapolis: University of Minnesota Press.

Mandel, E. 1975. *Late Capitalism*. London: Humanities Press.

Martin, W. 2009. Re-Programming Lyotard: From the Postmodern to the Posthuman Condition. *Parrhesia*, 8, 60–75.

Miller, J. 1983. *The American People and Science Policy: The Role of Public Attitudes in the Policy Process*. New York: Pergamon.

Moor, J.H. 2001. The Future of Computer Ethics: You Ain't Seen Nothin' Yet! *Ethics and Information Technology*, 3, 89–91.

Moor, J.H. 2005a. Should We Let Computers Get Under Our Skin?, in *The Impact Of The Internet On Our Moral Lives*, edited by R. Cavalier. Albany: State University of New York Press, 121–38.

Moor, J.H. 2005b. Why We Need Better Ethics for Emerging Technologies. *Ethics and Information Technology*, 7, 111–19.

Moore, G.E., 1965. Cramming More Components Onto Iintegrated Circuit. *Electronics*, 38 (8), 114–17.

Moravec, H. 1998. When Will Computer Hardware Match the Human Brain? *Journal of Evolution and Technology* [Online], Volume 1. Available at: http://www.transhumanist.com/volume1/moravec.htm [accessed: 5 December 2010].

Nuyen, A. 2004. Lyotard's Postmodern Ethics and Information Technology. *Ethics and Information Technology*, 6, 185–91.

Perri. 2001. Ethics, Regulation and the New Artificial Intelligence, Part I: Accountability and Power. *Information, Communication and Society*, 4 (2), 199–229.

Prewitt, K. 1983. Scientific Illiteracy and Democratic Theory. *Daedalus*, 112 (2), 49–64.

Sandywell, B. 2006. Monsters in Cyberspace Cyberphobia and Cultural Panic in the Information Age. *Information, Communication and Society*, 9 (1), 39–61.

Santillo, D. and Johnston, P. 1999. Is There a Role For Risk Assessment Within the Precautionary Legislation? *Human and Ecological Risk Assessment*, 5, 923–32.

Schulman, P.R. 1993. The Negotiated Order of Organizational Reliability. *Administration and Society*, 25, 353–72.

Sclove, R. E. 2010. *Reinventing Technology Assessment: A 21st Century Model*. Washington, DC: Science and Technology Innovation Program, Woodrow Wilson International Center for Scholars. Available at: http://www.wilsoncenter.org/techassessment [accessed: 5 December 2010].

Sharkey, N. 2009. The Robot Arm of the Law Grows Longer. *Computer*, 42 (8), 113–15.

Sim, S. 2001. *Lyotard and the Inhuman*. Cambridge: Icon Books.

Simpson, R. and de Garis, H. 1997. The Brain Builder. *Wired*. [Online]. Available at: http://www.wired.com/wired/archive/5.12/degaris.html [accessed: 5 December 2010].

Snook, S.A. 2000. *Friendly Fire: The Accidental Shootdown of U.S. Black Hawks over Northern Iraq*. Princeton: Princeton University Press.

Som, C., Hilty, L., and Ruddy, T.F. 2004. The Precautionary Principle in the Information Society. *Human and Ecological Risk Assessment*, 10, 787–99.

Sunstein, C. 2002. The Paralyzing Principle. *Regulation*, 25 (4), 32–7.

Tavani, H. 2001. The Current State of Computer Ethics as a Philosophical Field of Enquiry. *Ethics and Information Technology*, 3 (2), 97–108.

Tavani, H. 2007. *Ethics and Technology: Ethical Issues in an Age of Information and Communication Technology*. Boston: John Wiley and Sons.

Tickner, J., Kriebel, D. and Wright, S. 2003. A Compass for Health, Rethinking Precaution and its Role in Science and Public Health. *International Journal of Epidemiology*, 32, 489–92.

van den Belt, H. 2003. Debating the Precautionary Principle: 'Guilty until Proven Innocent' or 'Innocent until Proven Guilty?' *Plant Physiology*, 132, 1122–6.

Weckert, J. 2010. *In Defence of the Precautionary Principle*. IEEE International Symposium on Technology and Society (ISTAS), 7–9 June, 42–47.

Weizenbaum, J. 1984. *Computer Power and Human Reason: From Judgment to Calculation*. Harmondsworth: Penguin.

Whyte, J. 2007. Only a Reckless Mind Could Believe in Safety First. *The Times*, 27 July.

Winner, L. 1986. *The Whale and the Reactor: A Search for Limits in an Age of High Technology*. Chicago: University of Chicago Press.

Zimmerman, A.D. 1995. Toward a More Democratic Ethic of Technological Governance. *Science, Technology and Human Values*, 20 (1), 86–107.

5

Risk as Workers' Remembered Utility in the Late-Modern Economy[1]

Clive Smallman[2] and Andrew M. Robinson

Introduction

Unquestionably, the activities of mortgage lenders, banks, other financial advisors, financial services regulators and governments worldwide wrought havoc upon the world economy from December 2007 on, and this crisis continued to deepen as this book went to press. Layered over recent profound changes in economy and society, produced by 'globalization' and referred to as 'late-modernism', what we are witnessing is arguably the most uncertain, unpredictable and dynamic political, economic and social environment since the opening of the industrial age. When this in turn is overlaid with major global issues relating to climate change, population growth, the distribution of resources and the decline of biodiversity, then it is understandable why some scholars have chosen to paint a dark picture indeed for the future of humankind. For some, this is exemplified in the suffering of Asian and African peoples ruled by despots, crippled by natural disasters or disadvantaged by climate or a lack of resources. For others, writing in the organizational domain, there are similarly pernicious developments in late-modern workplace organization, and this is the subject of this chapter. There are of course countervailing positions, not least that some people may well be worried about global warming and

1   An earlier version of the this chapter was presented at the 2008 Annual Meeting of the Academy of Management, Anaheim, California, USA. The authors are grateful to the anonymous reviewers and meeting participants who were generous in their commentary before and at the meeting. We are also grateful to Simon Bennett for asking and encouraging us to make a contribution to this volume, and for his excellent critique of earlier versions.

2   Corresponding author: clive.smallman@lincoln.ac.nz

species attrition, but surely they are equally or more worried about keeping their job and creating a prosperous world for their offspring?

Whatever the nature of the changes and their effects, there is a consensus that they are being driven by two interrelated phenomena: *globalization* and the exponential growth of *information and communication technology* (ICT) innovation and diffusion. Bennett (2009) conforms to the conventional view of progressive globalization as a social process dating back to the nineteenth century with roots in the development of transport and communication technologies, the rapid growth of interdependent trade between Western Europe and the rest of the world, and foreign direct investment by European states into the non-industrialized world. In the last 25 years, we have seen an accelerated reversal of fortunes, with Asian countries, notably Japan and more recently China and India either making investments in Europe (to ease market entry) or purchasing 'dying' brands (e.g., Rover) and relocating industrial production to take advantage of cheap labour costs. In parallel Western corporations have sought to make cost savings by similarly relocating production or service functions throughout greater Asia. For some the travel of capital from West to East and *vice versa* is the result of:

> *innovating capitalist[s] … [e]nticed into creating new labour-saving technologies by the prospect of reaping higher rates of profit, these capitalists confront unavoidable competition and technology imitation which undermine their advantage and leads ultimately to falling prices and falling rates of profit. Their only recourse is to venture abroad, to globalize, to invest and produce in the less technologically developed economies. (Jellisen and Gottheil 2009)*

Yet, for those same authors:

> *This globalization of resource and product markets is a dual-edged sword: It creates and destroys. It destroys the economic and social fabrics of the less developed economies and creates in its wake a replica of the more technologically advanced. (Jellisen and Gottheil 2009)*

Nevertheless, it is the rapid development and remarkable penetration into our social and economic lives of ICT that has enabled or driven these changes (OECD 1989). The essential argument is that, largely through ICT, work is no longer nominally constrained by geography; instead workers are interconnected and workplaces are interwoven across the physical boundaries of workplaces and the legal boundaries of enterprises. The penetration of these technologies is such that they:

> ... *transform the nature, form and temporal aspects of work, influence the structure and meaning of the workplace and impact human relations in organizational settings ... They also lead to changes in ... power and authority [and] communication and participation .... (Gephart 2002)*

A simple case in point is the manner in which the present authors worked to develop this chapter (and indeed other publications). We are located as far apart as is possible temporally and geographically (12 hours and 19,000 kilometres or 12,000 miles). Yet, we are able to work asynchronously through the Internet and synchronously through either conventional telephony or voice over internet protocol (VOIP, e.g., Skype). The rare occasions on which we meet demands that one or the other of us travels for 30 plus hours. This works, but our relationship is somewhat different than when we both lived in Yorkshire in the north of England. Notably, our communication can easily become unsynchronized and, if we are honest, disharmony has occasionally crept in, when it might not have done so if we lived in the same time zone (so enabling a quick phone conversation). If we set these experiences in the context of a complex multinational corporation, then the veracity of Gephardt's statement is substantiated. However, a major issue for scholars attempting to understand how ICT-enabled globalization has affected the late-modern workplace is the 'macro-level' of the theory that has thus far been published. We contend that much of the literature conventionally cited in discussions of the late-modern workplace over-generalizes complex, interrelated or interwoven social and economic processes. Much of the published theory is polemic, little of it has an empirical base, and, arguably, even the most feted of theories (including Beck's *Risk Society*) suffer from problems of poor logic (Abbott, Jones and Quilgars 2005, Bulkeley 2001, Mythen 2005, Smith, Cebulla, Cox and Davies 2006, Strangleman 2007).

From this perspective, based on our ongoing work in workplace risk (Robinson and Smallman 2006) and in the context of this volume, our specific interest is in developing theory that enables us to properly evaluate risks posed to workers by workplace opportunities or threats in the late-modern economy. This has clear implications for the management of enterprise, especially in the context of three significant themes that are common across the relevant literature: the nature of jobs ('good' versus 'bad' – Coats 2006, Doeringer and Piore 1971); the form of employment arrangements (Kunda, Barley and

Evans 2002, Smith 2001); and the growth of high performance workplaces (Appelbaum, Bailey and Berg 2000, Whitfield 2000) grounded in the theory of flexible production (Hirst and Zeitlin 1991, Marshall 1999, Piore and Sabel 1984, Sabel 1989).

We develop a theoretical synthesis that draws upon sociology, institutional economics, employee relations and organizational behaviour, focused upon the eventual evaluation of evidence of heightened risk to employees in the 'brave new world of work'. We first define our understanding of risk and the late-modern economy, developing our research questions. From here we develop our theoretical framework, in order to identify the key propositions that underlie our questions. We discuss our thoughts on perceptions of opportunities for and threats to workers in the late-modern economy, reflecting upon personal experiences that provide partial answers to our questions. Our intended contribution is to the knowledge of *workplace risk management* through synthesizing a diverse range of theory, drawing together thinking around flexible specialization, the nature of jobs, and employment arrangements that significantly influence workplace risk.

Risks, Hazards, Opportunities and Threats

Risk is an ancient, pervasive and elemental part of life (Bernstein 1996); yet, perhaps the fundamental challenge that risk poses (especially to those who study it) is that it is always 'in the eye of the beholder'. However it is communicated, risk is ultimately an expression of an individual's socially constructed perception (Krimsky and Golding 1992) of the likelihood and impact of some event occurring. The common conceptualization of risk is overwhelmingly negative, associated with moral, social, economic or physical (natural or technological) hazards that take some tangible form (however transitory that may be); that is *threats*. However, an equally valid, if less common conceptualization is of risk associated with potential and tangible benefits (usually economic or political); that is *opportunities*. Such hazards or benefits are, for want of a better expression, *concrete*. However, the risk associated with a potential event is perceived differently by every individual exposed to it, and these perceptions change as they receive and process new information concerning the event (Klein 1998); risk may *concresce* (a temporary aggregation of perceived experiences), but it is *never* concrete. As such it is as fluid as the perceptions of the people exposed to potential events to which expressed or suppressed risk perception is one response. Hence, in this sense, all risk is *always* new; opportunities and threats though are somewhat different.

Natural, moral, economic and social hazards are well known to us, as are the effects of recently created technological hazards. The same is true of economic, political or social opportunities. However, over the past decade noted social theorists (Bauman 1998a, 1999, 2000, 2001, Beck 1992b, 1999, 2000, Giddens 1990) have each claimed to detect intensification in individual and collective *angst* about manufactured or anthropogenic hazards (Giddens 1999), implicitly or explicitly driven by globalization or ICT innovation and diffusion. The intensification of perceived risks associated with these hazards, their antecedents and their consequences is generally referred to as the *risk society* thesis, and has received widespread support (e.g., Benn, Brown and North-Samardzic 2009). The assessment and governance of risk in this context is regarded as challenging by many (e.g., Assmuth, Hildén and Benighaus 2010).

The risk society thesis predicts that in the workplace employees face increased distributive risks associated with their terms and conditions of employment, the trade-off between contractual flexibility and employment security, and the intensity and nature of work. By the same argument they also face increased political risks associated with the institutional costs of gains and losses of partnership between employers and employee representatives, the legitimacy of representation through unions, and the long-term basis of mutual trust (Martinez, Lucio and Stuart 2005).

The Late-Modern Economy

Before entering our discussion it is important to note that we adopt the convention 'late-modern economy' as specified by our editors, staying away from the issues and debate surrounding so-called 'post-modernism'. However, it is important to acknowledge, that much of the work we cite refers to the 'new economy'. We assume (always dangerous) that this equates to the 'late-modern' convention.

The origins of the late-modern economy lie in the movement away from mass production and standardization towards flexible specialization which characterized the mid-1980s (Piore and Sabel 1984) coupled to the economic boom in the West of the mid- to late-1990s. First defined (probably) by *Newsweek*, the late-modern economy first described a stable, service sector-based economy with a focus on wealth consumption, marking a move from an unstable, 'Fordist' production and wealth creating economy. Whilst the gains of the 1990s were lost in the 2001 recession, the concept endured, and it is now identified with contemporary approaches to the organization or management of enterprise in search of competitive advantage. In particular, the late-modern economy is strongly associated with the introduction of ICT in search of 'efficiencies' (increasing returns to the owners of capital who are increasingly short term in their outlook (Sennett 1998)), especially where that enables outsourcing of routine business functions. The late-modern economy is, to quote one of its more effusive supporters:

> ... *a knowledge- and idea-based economy where the key to higher standards of living and job creation is the extent to which innovative ideas and technologies are embedded in services, products, and manufacturing processes. It is an economy where risk, uncertainty, and constant change are the rule, rather than the exception. It is an economy where hierarchical organizations are being replaced by networked learning organizations. (New Economy Task Force 2000)*

However, in acknowledging the centrality of 'risk, uncertainty, and constant change' the New Economy Task Force (2000) also notes that 'many communities, community organizations, and individuals are simply unable to take advantage of' the late-modern economy. In addition to the social theorists cited earlier, for many authors, such inability is exacerbated or reflected in the technology driven shift in the nature of employment arrangements and of work (Ackerman *et al.* 1998, Andresky Fraser 2001, Appelbaum *et al.* 2000, Appelbaum and Batt 1994, Beder 2001, Blair and Kochan 2000, Cheever 2001, Ciulla 2000, Ehrenreich 2001, Freeman and Rogers 1999, Gorz 2000, Rifkin 1996, Sennett 1998, Smith 2001, Strangleman 2007, Zweig 2000). For the polemicists this change is 'for the worse' especially in employee relations, further promoting economic insecurity for low-wage workers, job insecurity for contingent labour, exclusion through gender or race, exploitation of minorities, and individualization of workers (away from representation through trade unions), and, at the extreme, loss of jobs (Aronowitz and Cutler 1998, Aronowitz and DiFazio 1994). For others, usually the more highly skilled, the late-modern economy apparently offers opportunity (Barley and Kunda 2006a).

Will the current recession affect this? We shall discuss this shortly, in the context of relevant literature.

Research Questions

In summary our research questions are: over recent time have there been significant changes in the locus of workplace power and authority, the nature of communication and participation, and the nature of jobs and employment arrangements; and have we seen increased adoption of so-called 'higher performance' work systems?

Relevant Literature

Building upon a notional post-Fordist second industrial divide (Gorz 2000, Piore and Sabel 1984) the risk society thesis identifies demands for a more mobile and flexible labour force, employment contracts and modes of production (Beck 1992b, Beck 1999, Allen and Henry 1997).

In preparing our theoretical synthesis, we first further define our *proxy* for workplace risk: remembered utility. We then explore theoretical and atheoretical perspectives on the nature of jobs and employment contracts, and then turn to explore theory focused upon flexibility in production.

Risk as Remembered Utility: Subjective and Workplace Wellbeing

We interpret distributive and political risks as *predicted utilities* (Kahneman 2000) in line with Kahneman, Wakker and Sarin's (1997) extension of Bentham's (1789/1948) notion of experienced utility as the experiences of pleasure and pain that 'point out what we ought to do, as well as determine what we shall do'. Such utilities are affective (Gilbert *et al.* 1998) or hedonic (Higgins 2006) forecasts of the likely outcome of some event (or set of circumstances) based upon perceptions grounded in an individual's life context (an acute expression of perceived risk). These perceptions in turn are defined as remembered utility (memory-based assessments of experienced utility in retrospective evaluations of past experiences (Kahneman 2000)). Memory-based experienced utility has been the almost exclusive focus of recent work in subjective wellbeing (Diener and Seligman 2004) in both the economic and psychology literatures. The derived and more narrowly defined concept of workplace wellbeing (a construct of physiological and psychological health) has also received considerable attention (Brown *et al.* 2006, Danna and Griffin 1999, Diener and Seligman 2004, Guest 2004, Helliwell 2006, Korpi 1997, Levy-Garboua and

Montmarquette 2004, Pastoriza, Ariño and Ricart 2008, Robinson and Smallman 2006, Twiname, Humphries and Kearins 2006, Vogt 2005).

At the socio-economic level work in remembered utility has focused almost exclusively upon subjective wellbeing (Diener 1984, Diener, Lucas and Scollon 2006, Diener and Seligman 2004, Diener, Suh and Oishi 1997). Based upon an individual, internal perspective the concept comprises of *satisfaction* with the various domains of life; *pleasant affect* associated with specific emotions such as joy, affection or pride; and *unpleasant affect*, which is associated with emotions or moods such as shame, guilt, sadness, anger or anxiety.

Conventional workplace wellbeing research has for many years focused upon the notion that happy workers produce good work, and this is broadly supported across the literature (Diener *et al.* 2004). Many significant antecedents of workplace wellbeing have been proven, but amongst the most frequently cited are: opportunity for personal control, opportunity for using skills, variety of tasks, physical security, supportive supervisor, respect and high status, interpersonal contact, good pay and fringe benefits, clear requirements and information on how to meet them (Diener and Seligman 2004, Warr 1999, Warr and Wall 1975). At a higher level of aggregation workplace wellbeing is the product of work situation, individual differences and occupational factors that give rise to physiological and psychological heath, which in turn has consequences for both the individual and the organization (Danna and Griffin 1999).

At both the societal and workplace levels, the research focuses upon self-declared judgments of how individuals feel about life and work, in other words self-declared remembered utility – a clear *proxy* for risk. In the context of this work we take that to be a function of job satisfaction, health and safety at work and general perceptions about work. Hence:

Proposition 1   If the risk society thesis is correct, then we would expect to see a significant decline in experienced utility over recent time.

'Good' Jobs, 'Bad' Jobs and Contingent Employment Arrangements: Institutionalism or 'Free Agency'?

The risk society thesis contends that the rise of technology in the workplace has resulted in contractual insecurity and dwindling employment opportunities (Beck 1998, Gorz 1982, 2000), to the extent that:

> *a steady, durable and continuous logically coherent and tightly-structured working career is ... no longer a widely available option. (Bauman 1998b)*

In this paradigm the insecurity of contingent work is associated with disruption in the temporal nature of work, with employers searching for efficiencies able to control contingent workers' hours in accordance with production or service requirements. In addition to the risk associated with the social and economic threats of variable hours of work, the thesis highlights issues associated with the decline and fragmentation of collective bargaining and transfer of employment costs to individual workers from employers. The thesis also asserts the need for an evolution in the configuration of employee relations (Mythen 2005), in order to help workers mitigate the risk associated with economic and social threats associated with the 'instability, volatility, flexibility, ephemerality and insubstantiality' (Gorz 2000) that defines the late-modern economy. Again, we turn to Bauman (1998a) for a succinct summary:

> *The pressure today is to dismantle the habits of permanent, round-the-clock, steady and regular work; what else may the slogan of 'flexible labour' mean? The strategy commended is to make the labourers forget, not to learn, whatever the work ethic in the halcyon days of modern industry was meant to teach them. Labour can conceivably become truly 'flexible' only if present and prospective employees lose their trained habits of day-in-day-out work, daily shifts, a permanent workplace and steady workmates' company; only if they do not become habituated to any job, and most certainly only if they abstain from (or are prevented from) developing vocational attitudes to any job currently performed ...*

We argue that there are strong parallels between the risk society thesis and dual labour market theory (Baron and Bielby 1984, Berger and Piore 1980, Bosanquet and Doeringer 1973, Doeringer and Piore 1971, Hudson 2007, Osterman 1984, Piore and Sabel 1984, Tilly 1996), which highlights work inequality in its distinction between primary jobs and secondary jobs (Coats 2006). The primary sector provides good remuneration, good working conditions, stable employment, careers, job security, equity and other benefits. The secondary sector is characterized by short-term employment relationships, little or no prospect of internal promotion, and the determination of wages primarily by market forces. In terms of occupations, it consists primarily of low or unskilled jobs. Whether they are manual or administrative, they are characterized by low skill levels, low earnings, easy entry, job impermanence, and low returns to education or experience (Barley and Kunda 2006a, Bone 2006, Doeringer and Piore 1971, Hudson 2007). Like the risk society theorists, 'institutionalists' (Kalleberg *et al*. 1997, Osterman 1988, Parker 1994, Smith 1998) employ dual labour market theory to warn that the growth of contingent work (secondary labour) will not only undermine the primary sector (promoting job insecurity (Belan, Carré and Gregoir 2010)), but also increase demand for social support for the less advantaged members of society or minorities (Cohen and Haberfield 1993, Hipple and Stewart 1996, Polivka 1996). The migration of cross-border service workers in Europe is a pressing case in point (Dølvik and Visser 2009). The institutionalists also argue against the shift to contingent labour on the grounds that this would fragment the collective power of workers by emasculating trade unions (Aronowitz and DiFazio 1994, Rifkin 1996) and this is borne out by research into the future of workforce representation (Healy *et al*. 2004). However, caution is required in adopting this universal perspective. Recent research suggests that in the USA the women's labour market does not exhibit the conventional dual labour market segmentation observed in the men's labour market (Meyer and Mukerjee 2007). Instead they define 'high wage' and 'low wage' work, which they find to be strongly related to race and to human capital formation:

> *In the high-wage sector women, regardless of race, experience significant returns to general and firm-specific human capital formation. These skills, however, depreciate rapidly if the woman leaves the labour force. In contrast, in the low wage sector, time out of the labour force does not seem to indicate a depreciation of human capital for women of either race. Racial differences do show up however, in the low wage sector in rewards to general training. Race is a disadvantage for the black woman in this sector, but she can overcome this disadvantage through education and uninterrupted job tenure. On the other hand, her white counterpart only receives returns to job-specific skills. (Meyer and Mukerjee 2007)*

A counter to these critical theories of heightened perceived risk to the labour force lies, in the manner of the conventional debate between labour and capital, with the 'New Right' proponents of a market-orientated approach to coping with the unraveling of secure work and its supporting institutions (Marshall, 1999). The 'free-marketeers' focus mainly on contracting, a very specific form of contingent employment, advocating its role in a paradoxical libertarian, anti-corporate rebellion (Barley *et al*. 2006a, Barley and Kunda 2006b, Beck 1992a, Bridges 2001, Kunda *et al*. 2002, McGovern and Russell 2001, Pink 2001, Reinhold 2001). Implicitly taking their lead from the same post-Fordist cues as the institutionalists they:

> *... portrayed corporate life as stifling and petty ... [which] ... forced people to play 'politics' and subject themselves to the whims of incompetent managers for inadequate pay ... [adding that] ... 'jobs' and 'careers' were outmoded inventions of the industrial revolution, designed for the benefit of employers. (Barley and Kunda 2006a)*

The free-marketeers promote 'emancipation' as a response to the demise of traditional employment, refusing loyalty and seeking freedom to develop and market their knowledge, experience and skills to any organization (Barley and Kunda 2006b, Kunda *et al.* 2002). They also view fixed term contracts as a means of coping with demand fluctuations and as a way of securing the employment of the core workforce (e.g., Pfeifer 2009). This liberal vision of economic individualism runs contrary to the despair implicit in the argument of the risk society and dual labour market theories. Free-marketeers argue that contingency is about 'liberation rather than isolation'; that it minimizes 'uncertainty about employment'; that it enhances flexibility and personal control; that it is about pay that reflects the real value of skills; and that it promotes 'self-actualization rather than estrangement' (Barley and Kunda 2006a).

There are difficulties with both the institutional and free-market perspectives. Early work in institutionalism was criticized for being atheoretical (Dickens and Lang 1988), but subsequent work (Baffoe-Bonnie 2003, Bulow and Summers 1986, Heckman and Hotz 1986, Hirsch 1980, Kleven, Kreiner and Dixon 2002, McDonald and Solow 1985, Piore 1975) has largely offset this arguable fault. However, the socio-economic theoretical base is biased towards lower-skilled temporary workers, whereas free-market *atheoretical* prescriptions are based largely upon anecdote and the professional experiences of their advocates and their professional networks. Even allowing that this 'hearsay' evidence has validity, it is problematic because it focuses almost exclusively upon highly skilled contingent professionals. Neither perspective accommodates the range of contingent employment and as a consequence:

> *institutionalists [conflate] the effects of contingent employment with correlates of low-skilled work and advocates of free agency [peddle] images of contracting built solely on the experiences of an elite. (Barley and Kunda 2006a)*

Exacerbating, as well as explaining, their common lack of coverage of all forms of contingent work is the ideological grounding of both perspectives. As is the case of proponents of free-markets in all disciplines the market for contingent employment is seen as a panacea for the decaying system of modern work. For institutionalists, the issue is one of emancipating low-skilled contingent workers from oppression and exploitation by self-interested employers encouraged by governments lacking the ability or willingness to legislate for and enforce employment protection (Morgan 1998).

In an interesting counterpoint to the Western experience, and in line with the globalization argument, Zhang (2008) finds that two different models of labour controls have recently emerged in the Chinese automotive industry. In one, the *lean-and-dual* model, management adopts labour force dualism, intermingles formal contract workers with agency workers on production lines. This hybrid workplace arrangement so combines both *hegemonic* and *despotic* industrial relations, with the former based on high wages, generous benefits, better working conditions, and relatively secure employment for formal workers. Despotic labour relations characteristically offer poor conditions for temporary agency workers with lower wages and insecure employment. In the alternative model, *lean-and-mean* automotive firms adopt an insecure, high-wage, high-turnover strategy of lean production for their entire workforce. This is particularly interesting in the context of Jellisen and Gottheil's (2009) Marxist perspective on globalization:

> In the end, we are all one global economy, western in character. Marx sees this inevitable globalization process as progressive and praiseworthy.

Hence:

Proposition 2   Workers' terms and conditions of employment are significantly associated with experienced utility. (Martinez Lucio and Stuart 2005)

Flexible Production and 'High Performance' Work

As a strand of 'supply-side institutionalism' (Streeck 1991) the theory of 'flexible specialization' (Piore and Sabel 1984, Sabel 1989) aspires to superior competitive advantage for firms through the application of ICT, allowing the development of diversified product portfolios and 'non-price competition policies' (Marshall 1999). This it claims can be achieved in combination with 'high wages, skilled labour and flexible non-'Taylorist' organization of work'. In other words this reverses the so-called 'Fordist' trend of fragmented and under-skilled labour, allowing the development of 'better work' in which versatile and highly skilled workers are 'empowered' often in self-managed teams. This analysis has been widely influential in the USA and in Britain (Marshall 1999), although has not been without its critics (Elam 1990, Williams *et al.* 1987), particularly in early expositions, when its ambiguity and lack of theoretical clarity were questioned. Moreover, it is arguable that whilst senior management may have ceded production control to front-line workers, strategic power has become much more concentrated in the hands of management. Thus workers are faced with implementing strategies, in the formulation of which they have almost no control – perhaps transferring the risk of failure onto the workforce?

Certainly in the USA and Britain, as well as some other 'Western' countries, current forms of work organization exemplify a number of trends related to the development of workplace aspects of flexible specialization: the reduction in job demarcation through functional flexibility facilitated by multi-skilling, enabling workers to undertake a wide variety of tasks within the workplace, acting in either implicit or explicit partnership with management (Richardson *et al.* 2005). Under this approach, defined loosely as 'high performance work systems', workers are often now regarded as 'skilled problem solvers' who think more holistically, autonomously deciding how to undertake their work tasks (Whitfield 2000), that is they have greater influence over their own work. It is arguable that devolving control to workers is a tacit reassertion of work intensification and managerial control (Gallie, Felstead and Green 2001). However, proponents of this approach argue that by placing a higher value on employee knowledge and skills, employers offer 'empowered' workers a more challenging and fulfilling work environment (Smith 1997). Hence:

> Proposition 3   In encountering the introduction of flexible production or high performance work systems over recent time workers have 'enjoyed' increased experienced utility.

So, what does the future hold?

The Evolution of the Economy, the Labour Market and the Future of Work? 2008 Onwards

Any number of official sources painted a bleak picture of the economy and the labour market following the 2008 recession (Office for National Statistics 2009,

2010a, 2010b, 2010c). Independent sources too predicted 'lowered employment intentions', i.e., increased redundancies for the third quarter of 2010 (CIPD/ KPMG 2010). Jenkins (2010) suggests that this is a normal pattern in the wake of a recession, and that the trend is similar to those following the recession of the 1980s and 1990s. However, he also notes that the labour market has been more resilient this time (Jenkins (2010)). So, if we are following a nominal pattern of employment following a recession, what is the future of work? This subject has exercised authorities for some time now (e.g., Malone (2004), Donkin (2010)), and there are very many different versions at many different levels of abstraction.

Using their immense global reach, PricewaterhouseCoopers (2007, 2008, 2009, 2010) paint what we concur is a realistic set of future worlds of work. Based upon 'global forces that will have significant influence' (PricewaterhouseCoopers 2007), i.e., individualism versus collectivism, and corporate integration version fragmentation, PwC (2007) identify three co-existent scenarios for the world of work.

In the *orange* world 'small is beautiful' with a balance struck between collectivism and individualism, but with a bias to corporate fragmentation. In this world:

> *companies begin to breakdown into collaborative networks of smaller organizations; specialization dominates the world economy. (PricewaterhouseCoopers 2007)*

With the fragmentation of global businesses, localism comes to the fore and a 'low impact, high-tech' business model is empowered by technology (e.g., in supply chains and to develop social capital and collaboration). Fragmentation leads major corporations to fail and external networks are key to enabling flexible and adaptive smaller companies. Individuals take sole responsibility for skills development and view themselves as members of a skill or professional network rather than as an employee of a specific company. Contract employees are hired based on their skill sets (PricewaterhouseCoopers 2007).

The impact of the recession on the orange world is forecast in five specific areas: a shift of focus away from innovation and risk taking; cost cutting; lessened technological investment; brand damage through vexatious electronic 'vandalism' (on networking sites); and a loss of interest by millennials in companies who do not maintain their technological 'edge'. (PricewaterhouseCoopers 2009)

There is clear resonance of the orange world in both of the latter two literatures we discussed earlier, especially in terms of contingent employment arrangements, and high performance work. As PwC (2007) note, this is a radical world of work. The orange world epitomizes contingent work arrangements and requires high performance. Yet as the counter-argument posits, it also marginalizes a huge number of the working population.

Where collectivism dominates in an integrated corporate world, 'companies care' in a green world of work and:

> social responsibility dominates the corporate agenda with concerns about demographic changes, climate and sustainability becoming the key drivers of business. (PricewaterhouseCoopers 2007)

In this world companies are characterized by a 'green sense of responsibility' as well as a 'powerful social conscience'. This is because environmental performance and business ethics are primary customer requirements. This forces strong alignment between business and society (PricewaterhouseCoopers 2007). Organizations are highly regulated, with particularly stringent employment law. As a consequence, risk management and quality assurance are particularly vital functions, but so too are personnel development functions and the importance of individual social contributions. In essence the corporate responsibility agenda is combined with people management (PricewaterhouseCoopers 2007).

The forecast effects of the 2008 recession on the green world (PricewaterhouseCoopers 2009) suggest that companies must exercise caution in radically changing reward and bonus systems. The forecast shortage of talent in a green world means that people will go elsewhere if they feel they are not being rewarded. Creating a highly risk averse culture – a natural reaction to the events of 2008 onwards – too is ill advised, since it is forecast to 'hamper creativity, innovation and profitability over the long term'. It may also affect speed of delivery. Pragmatically ignoring sustainability or corporate responsibility as 'non-critical' is also ill advised, with the potential for long-term brand damage. Global organizations rely on social capital for success, which requires face-to-face contact on a regular basis. Cost cutting by limiting travel puts these networks at risk. Technology is an alternative, but not all of the time.

Perhaps the most challenging implication of the downturn for organizations operating in the green world is disillusionment with them. Countering these requires careful engagement with employees and stakeholders.

In one sense the green world is a predictable response to the risk society thesis and more so following the recession. It emphasizes regulation, quality assurance and risk management, and employment security. But how realistic is it?

In the blue world 'corporate is king' and:

> big company capitalism rules as organizations continue to grow bigger and individual preferences trump beliefs about social responsibility. (PricewaterhouseCoopers 2007)

Herein corporate careers are highly sought after because of excellent training and excellent reward packages. The consumer is dominant and globalization is the defining corporate trend. The gap between rich and poor widens. In this world *Fordism persists* because of the need to create consistency across the supply chain in pursuit of quality assurance across large-size and wide-scale operations. There is careful career and education management, but with greater responsibility taken by individuals for the management of their careers. Outside of the corporate sphere, individual development and financial benefits are rarely as good (PricewaterhouseCoopers 2007).

In the blue world, the impact of the recession will be most evident in the costs' base of corporations. Here PwC (2009) see threats in the workplace relating to the reduction of graduate intake numbers – bad for graduates in the short term, but worse for corporations in the longer term. It is also likely that training and development budgets will be cut. Not only can this affect quality or service, it will threaten the company's ability to recover when the upturn arrives, and their reputation as a good employer. The propensity of employees to commit fraud also increases in a downturn, exacerbating the effects of a downturn caused by poor managerial behaviours. Fraud is often attributed to poor morale and personal financial pressures.

Looking forward from these scenarios and based on a survey of graduates' expectation of work, PwC (2008) forecast a 'talent crunch' because 'people supply will be the most critical factor for business success'. This means that the millennials (those born around the year 2000) will be a privileged generation of workers because those with the right skills will be in high demand attracting lucrative reward packages. There is a forecast shift in the employment relationship with power ceding to employees. Millennials expect job mobility (this is affirmed strongly in PwC (2010)), but at the same time require corporate responsibility. They will not, counter to some predictions, reject conventional work practices with office-based work expected and home-working an exception. Regular office hours are expected with only a minority expecting flexible hours. The millennials value training and development highly. They expect to fund their own retirement.

Conclusion

Our synthesized theory is that high performance work systems and contingent employment (both as outcomes of the late-modern economy) are significantly associated over time with distributional and political risks (expressed as utility) to employees. As we have previously demonstrated, the direction of that association is disputed between institutionalists (including risk society theorists) on the one hand and free-marketeers on the other. Both groups have attempted to empirically validate their claims. With the possible exception of Smith (2001) no member of either group has attempted a complete coverage of primary and secondary labour markets, instead apparently preferring samples that support their contentions. PwC's (2007, 2008, 2009, 2010) futures of work too suggest changes in work systems and employment arrangements, whilst implicitly recognizing the 'rise of risk'.

Parts of this chapter were written in late 2007, and it is fair to say that the economic landscape has shifted somewhat since then. In 2007, few had any inkling of the cataclysm that was approaching (as reflected in PwC (2009), Office of National Statistics (2009, 2010a, 2010b 2010c), CIPD/KPMG (2010) and Jenkins (2010)). Those that did (Clive was one) had either researched or worked in operational risk or another aspect of financial services, witnessing the culture of excess, pressure on costs (minimizing the wages of 'normal' workers who supported the high earners), disregard for all aspects of business risk and the dilution or ignorance of regulation that was prevalent not just in the City of London (Bennett 2009), but across the USA, Europe, Asia and Australasia too.

Looking further back to 2001 when Clive worked in financial services, first in London then in Cardiff, there were signs then, particularly in London that all was not well. The physical and social environment of London was at that time a patina of respectability. Once outside the sheen of the Square Mile of the City, particularly to the east, the back streets were often dirty, poorly lit and frequented by prostitutes (male and female). Walking home to his rented accommodation was never short of 'entertainment' or risk. Outside of the well-to-do traders and bankers and the highly paid information technology contractors (of which Clive was one), the income gap to service workers was enormous. The gap to the next layer was painfully obvious.

Was there a dual labour market? Without doubt, but it exhibited both sides of the dialectic. Clive was a high earning contractor, but with a role that could be terminated at any time. Insecurity in this sense is a contractor's lot. On the other hand on a daily basis we saw evidence of those employed in 'contracted out' services (such as refuse collection), whose jobs are vital to the City, but who lived a pressured, dirty existence, and there was evidence of many jobs such as these. In this sense there exists a multilayered labour market with the super rich, the high earning contracting or consulting, the 'respectable' service work (teacher, emergency services), the less respected work, and the 'rest' (those poor souls with nothing, who sleep in cardboard boxes). The talk of cardboard boxes takes us to a metaphor for the 'dark side' of late-modern city life inspired by Bennett (2009).

In Chicago in August 2009, Clive observed at first hand Wacker Drive, which encapsulates graphically the gloss of late-modern America: a wide and clean boulevard bounded on one side by the Chicago River and to the other by monumental buildings of steel, stone and glass. Between the river and the buildings are wide pavements. Along those pavements, even in the fetid heat of an Illinois summer, walk the 'beautiful people'. These immaculately presented service workers parade wealth, even in such hard times. In and amongst them walk more ordinary folk, less well presented, but obviously getting by in spite of the times. There are also 'interlopers', asleep on benches (not for long if the police spot them) or begging for a few dollars to buy alcohol or worse. They too are inhabitants of Wacker Drive, just not the level we are walking upon, for this is *Upper* Wacker Drive. The Chicago River does not run at the level of Upper Wacker, but a storey below, for Wacker Drive is 'double-decked' (in so many ways) with what might be described as a 'broken' boulevard running under its richer sibling. At any time of the day, but particularly at night, *Lower* Wacker Drive is Upper Wacker's dark side. Lower Wacker is the definitive second tier. In too many places it 'houses' the dispossessed and dissolute (often in cardboard boxes), it is polluted, it is filthy, it is risky and dangerous and it is certainly not a place to be at any time of the day, and definitely not at night. Wacker Drive then is a tellingly accurate metaphor for late-modernism as constructed by many scholars

Back in the City of London in 2001, was there evidence of a risk society? Again without doubt, as Clive vividly recalls standing in a pub in the City watching with disbelief, along with colleagues, the events unfolding in New York on 11 September 2001. As the World Trade Center was first attacked, burned and then fell, consuming 2,995 people, it was apparent the world had changed and that socio-political risk was on the rise. Those dramatic events aside, contractors live with risk; they always have and they always will. It is part of the deal when you 'sign up' for this type of work, but it is well paid, and you can develop a lifestyle that suits you (one colleague worked for six months of the year and spent the remainder of the year at his apartment in Spain). For the workers on contracted out refuse collection and those jobs that actually permit the City to function are not so fortunate. 'Efficiency' is the word that governs their existence, and in effect is a synonym for risk.

More specifically, what of our propositions? Is there evidence of a significant decline in experienced utility over time? In the context of the recession and its aftermath, for the majority of people (in work and out of work) the answer is almost certainly 'yes'. But, for a few that 'dark cloud' has 'come bearing gifts' (forgive the mixed metaphor). However, this is 'normal' in a recession and is arguably a natural part of the economic process. In short a more accurate answer is 'it depends'.

What of the association between experienced utility and workers' terms and conditions of employment? As with the previous proposition, the 'answer' – on the basis of what we know now – depends upon the worker concerned. For those who come from a strong trade or labour union tradition or those who espouse the benefits of a neo-Marxist approach to workplace organization, the answer seems to be that workers are at considerably increased risk. Others relish and exploit what they regard as opportunities in a freer labour market.

Finally, what of the effects of flexible production on experienced utility? Again, impact is in the eye of the beholder. There is a significant body of work that claims workers' rights for 'safe' employment have been overturned. Yet there are other counter-examples where individuals relish and take up the challenges posed by 'lean production'.

At the outset of this chapter, we espoused the need for a theory that would enable us to evaluate workplace risk. As one of our earlier reviewers commented, 'you have tried to combine too many theories, and made it all too complicated', and as is conventional for reviewers striving for so-called 'scientific objectivity', thereby simultaneously 'got' and missed the point. As our ambiguous answers to our propositions suggest, what we have here, we believe is the beginning of a theory, but one that needs careful exploration, needing considered study of workplace processes, hazards and the risks perceived by workers within them. Previous empirical work has focused largely upon developing a quantitative understanding of the various issues we have covered above, particularly in the areas of terms and conditions of employment, employment arrangements, partnerships, employment flexibility and trust. There has of course also been valuable case study work, exploring causation. Valuable normative work has also developed important concepts, notably around the area of distributive and political risks. However, there seems to have been little or no empirical work undertaken that looks at the development of workplace risk over time, and certainly not in the context of the risk society thesis or dual labour market theory.

The next step, we believe, is to develop a strand of ethnographic work that follows the practice of work. There is a rich, if narrowly focused research tradition here, exemplified in the work of Bennett (2010), Purser (2006) and Hodson (2002). Turner (1995) and Darrah (1996) are models of lengthier treatments. Zickar and Carter (2010) elegantly echo our call:

> We chronicle the early history of organizational research in which ethnography was an important methodological tool used to study workers' experiences. Early applied psychologists and management researchers were conversant with leading ethnographies and cited their work, occasionally even doing ethnography themselves. Although there is currently a vibrant niche of organizational researchers who use ethnography, the vast majority of organizational researchers have relied less on workplace ethnographies, citing them infrequently. We outline the benefits of ethnography and explain reasons why organizational researchers should reconnect with the spirit of ethnography, even if practical constraints keep them from conducting ethnographical work themselves.

This resonates strongly with us, and we endorse the need to develop a considered exploration of workers' perceptions of risk as we move beyond the late-modern workplace that is being constructed as we write and as you read. However, we would go further than the development of conventional ethnographies, seeking to develop properly 'engaged scholarship' (Van de Ven 2007) that is useful to practice (Albers Mohrman, Gibson and Mohrman 2001). What we seek is a detailed and rich understanding of employment practices and processes and workplace routines in late-modern workplaces, with a particular focus upon risk as workers remembered utility.

References

Abbott, D., Jones, A. and Quilgars, D. 2005. *Reflecting on the Risk Society: Does Difference Make a Difference?* Paper presented at the Learning about Risk: ESRC Social Contexts and Responses to Risk Network Conference, Canterbury Cathedral International Study Centre.

Ackerman, F., Goodwin, N.R., Dougherty, L. and Galagher, K. (eds) 1998. *The Changing Nature of Work*. Washington, DC: Island Press.

Allen, J. and Henry, N. 1997. Ulrich Beck's 'Risk society' at work: Labour and employment in the contract service industries. *Transactions of the Institute of British Geographers*, 22(2), 180–96.

Albers Mohrman, S., Gibson, C.B. and Mohrman, A.M. 2001. Doing research that is useful to practice: A model and empirical exploration. *Academy of Management Journal*, 44(2), 357–75.

Andresky-Fraser, J. 2001. *White-Collar Sweatshop: The Deterioration of Work and Rewards in Corporate America*. New York: Norton.

Appelbaum, E., Bailey, T. and Berg, P. 2000. *Manufacturing Advantage: Why High Performance Work Systems Pay Off*. London: Cornell University Press.

Appelbaum, E. and Batt, R. 1994. *The New American Workplace: Transforming Work Systems in the United States*. New York: ILR Press.

Aronowitz, S. and Cutler, J. (eds) 1998. *Post-Work: the Wages of Cybernation*. London: Routledge.

Aronowitz, S. and DiFazio, W. 1994. *The Jobless Future: Sci-Tech and the Dogma of Work*. Minneapolis: University of Minneapolis.

Assmuth, T., Hildén, M. and Benighaus, C. 2010. Integrated risk assessment and risk governance as socio-political phenomena: A synthetic view of the challenges. *Science of the Total Environment*, in press.

Baffoe-Bonnie, J. 2003. Distributional assumptions and a test of the dual labour market hypothesis. *Empirical Economics*, 28(3), 461–78.

Barley, S.R. and Kunda, G. 2006a. Contracting: A new form of professional practice. *Academy of Management Perspectives*, 2(1), 45–66.

Barley, S.R. and Kunda, G. 2006b. *Gurus, Hired Guns, and Warm Bodies: Itinerant Experts in a Knowledge Economy*. Princeton, NJ: Princeton University Press.

Baron, J.N. and Bielby, W.T. 1984. The organization of work in a segmented economy. *American Sociological Review*, 49, 454–73.

Bauman, Z. 1998a. *Globalization: The Human Consequences*. Cambridge: Polity Press.

Bauman, Z. 1998b. *Work, Consumerism and the New Poor*. Buckingham: Open University Press.

Bauman, Z. 1999. *In Search of Politics*. Cambridge: Polity Press.

Bauman, Z. 2000. *Liquid Modernity*. Cambridge: Polity Press.

Bauman, Z. 2001. *The Individualized Society*. Cambridge: Polity Press.

Beck, N. 1992a. *Shifting Gears: Thriving in the New Economy*. Toronto: Harper Collins.

Beck, U. 1992b. *Risk Society: Towards a New Modernity*, translated by M. Ritter. London: Sage.

Beck, U. 1998. *Democracy Without Enemies*. Cambridge: Polity Press.

Beck, U. 1999. *World Risk Society*. Cambridge: Polity Press.

Beck, U. 2000. *The Brave New World of Work*, translated by P. Camiller. Cambridge: Polity Press.

Beder, S. 2001. *Selling the Work Ethic: from Puritan Pulpit to Corporate PR*. London: Zed Books.

Belan, P., Carré, M. and Gregoir, S. 2010. Subsidizing low-skilled jobs in a dual labour market. *Labour Economics*, in press.

Benn, S., Brown, P. and North-Samardzic, A. 2009. A commentary on decision-making and organisational legitimacy in the Risk Society. *Journal of Environmental Management*, 90(4), 1655–62.

Bennett, S. 2009. Londonland. *An Ethnography of Labour in a World City*. London: Middlesex University Press.

Bennett, S. 2010. A longitudinal ethnographic study of night-freight pilots. *Journal of Risk Research*, 13(6), 701–30.

Bentham, J. 1789/1948. *An Introduction to the Principle of Morals and Legislations*. Oxford: Blackwell.

Berger, S. and Piore, M.J. 1980. *Dualism and Discontinuity in Industrial Societies*. New York: Cambridge University Press.

Bernstein, P.L. 1996. *Against the Gods: The Remarkable Story of Risk*. Chichester: John Wiley and Sons Inc.

Blair, M.M. and Kochan, T.A. (eds) 2000. *The New Relationship: Human Capital in the American Corporation*. Washington, DC: Brookings Institution Press.

Bone, J. 2006. 'The longest day': 'Flexible' contracts, performance-related pay and risk shifting in the UK direct selling sector. *Work, Employment and Society*, 20(1), 109–27.

Bosanquet, N. and Doeringer, P.B. 1973. Is there a dual labour market in Great Britain? *The Economic Journal*, 83(330), 421–35.

Bridges, W. 2001. *Job Shift: How to Prosper in a Workplace Without Jobs*. Reading, MA: Addison Wesley.

Brown, A., Charlwood, A., Forde, C. and Spencer, D. 2006. *Changing Job Quality in Great Britain 1998–2004*. Employment Relations Research Series (WERS 2004 Grants Fund). London: Department of Trade and Industry.

Bulkeley, H. 2001. Governing climate change: The politics of the risk society. *Transactions of the Institute of British Geographers*, 26, 430–47.

Bulow, J.I. and Summers, L.H. 1986. A theory of dual labour markets with application to industrial policy, discrimination and Keynesian unemployment. *Journal of Labour Economics*, 4(3), 376–414.

Cheever, B. 2001. *Selling Ben Cheever: Back to Square One in a Service Economy*. New York: Bloomsbury.

CIPD/KPMG. 2010. *Labour Market Output: Quarterly Survey Report, Summer 2010*. London: Chartered Institute of Personnel and Development. Available at: http://www.cipd.co.uk/NR/rdonlyres/DE1D59DA-2424-4ABC-931E-ADDC8CC87329/0/5302_LMO_report_Summer10.pdf [accessed: 22 September 2010].

Ciulla, J.B. 2000. *The Working Life*. New York, NY: Three Rivers Press.

Coats, D. 2006. No going back to the 1970s? The case for a revival of industrial democracy. *Public Policy Research*, 13(4), 262–71.

Cohen, Y. and Haberfield, Y. 1993. Temporary help service workers: employment characteristics and wage determinants. *Industrial Relations*, 32, 272–87.

Danna, K. and Griffin, R.W. 1999. Health and wellbeing in the workplace: A review and synthesis of the literature. *Journal of Management*, 25(3), 357–84.

Darrah, C.N. 1996. *Learning and Work: an Exploration in Industrial Ethnography*. New York, NY: Garland Publishing.

Dickens, W.T. and Lang, K. 1988. The re-emergence of segmented labour theory. *The American Economic Review*, 78(2), 129–34.

Diener, E. 1984. Subjective wellbeing. *Psychological Bulletin*, 95(3), 542–75.

Diener, E., Lucas, R. and Scollon, C.N. 2006. Beyond the hedonic treadmill: Revising the adaptation theory of wellbeing. *American Psychologist*, 61, 305–14.

Diener, E. and Seligman, M.E.P. 2004. Beyond money: Toward an economy of wellbeing. *Psychological Science in the Public Interest*, 5(1), 1–31.

Diener, E., Suh, E. and Oishi, S. 1997. Recent findings on subjective wellbeing. *Indian Journal of Clinical Psychology*, 24(1), 25–41.

Doeringer, P.B. and Piore, M.J. 1971. *Internal Labour Markets and Manpower Analysis*. Lexington, MA: D.C. Heath and Company.

Dølvik, J.E. and Visser, J. 2009. Free movement, equal treatment and workers' rights: Can the European Union solve its trilemma of fundamental principles? *Industrial Relations Journal*, 40(6), 491–509.

Donkin, R. 2010. *The Future of Work*. Basingstoke: Palgrave-Macmillan.

Ehrenreich, B. 2001. *Nickel and Dimed: On (Not) Getting by in America*. New York: Henry Holt.

Elam, M.J. 1990. Puzzling out the post-Fordist debate: Technology, markets and institutions. *Economic and Industrial Democracy*, 11(1), 9–37.

Freeman, R.B. and Rogers, J. 1999. *What Workers Want*. Ithaca, NY: Cornell University Press.

Gallie, D., Felstead, A. and Green, F. 2001. Employer policies and organizational commitment in Britain 1992–1997. *Journal of Management Studies*, 38(8), 1081–101.

Gephart, R.P. 2002. Introduction to the brave new workplace: Organizational behaviour in the electronic age. *Journal of Organizational Behaviour*, 23, 327–44.

Giddens, A. 1990. *The Consequences of Modernity*. Cambridge: Polity Press.

Giddens, A. 1999. Risk and responsibility. *Modern Law Review*, 62(1), 1–10.

Gilbert, D.T., Pinel, E.C., Wilson, T.D., Blumberg, S.J. and Wheatley, T.P. 1998. Immune neglect: A source of durability bias in affective forecasting. *Journal of Personality and Social Psychology*, 75(3), 617–38.

Gorz, A. 1982. *Farewell to the Working Class: An Essay on Post-Industrial Socialism*. London: Pluto Press.

Gorz, A. 2000. *Reclaiming Work: Beyond the Wage-Based Society*. Cambridge: Polity Press.

Guest, D.E. 2004. Exploring the paradox of unionized worker dissatisfaction. *Industrial Relations Journal*, 35(2), 102–21.

Healy, G., Heery, E., Taylor, P. and Brown, W. 2004. *The Future of Worker Representation*. Basingstoke: Palgrave Macmillan.

Heckman, J.J. and Hotz, V.J. 1986. An investigation of the labour market earnings of Panamanian males: evaluating the sources of inequality. *Journal of Human Resources*, 21(4), 507–42.

Helliwell, J.F. 2006. Wellbeing, social capital and public policy: what's new? *Economic Journal*, 116, C34–45.

Higgins, E.T. 2006. Value from hedonic experience and engagement. *Psychological Review*, 113(3), 439–60.

Hipple, S.F. and Stewart, J. 1996. Earnings and benefits of workers in alternative work arrangements. *Monthly Labour Review*, 119, 46–54.

Hirsch, E. 1980. Dual labour market theory: A sociological critique. *Sociological Inquiry*, 50(2), 133–45.

Hirst, P. and Zeitlin, J. 1991. Flexible specialization versus post-Fordism: Theory, evidence and policy implications. *Economy and Society*, 20(1), 1–56.

Hodson, R. 2002. Worker participation and teams: New evidence from analyzing organisational ethnographies. *Economic and Industrial Democracy*, 23(4), 491–528.

Hudson, K. 2007. The new labour market segmentation: Labour market dualism in the new economy. *Social Science Research*, 36, 286–312.

Jellissen, S.M. and Gottheil, F.M. 2009. Marx and Engels: In praise of globalization. *Contributions to Political Economy*, 28(1), 35–46.

Jenkins, J. 2010. The labour market in the 1980s, 1990s and 2008/09 recessions. *Economic and Labour Market Review*, 4(8), 29–36.

Kahneman, D. 2000. Experienced utility and objective happiness: A moment-based approach, in *Choices, Values and Frames*, edited by D. Kahneman, and A. Tversky. Cambridge: Russel Sage Foundation and Cambridge University Press, 673–92.

Kahneman, D., Wakker, P.P. and Sarin, R. 1997. Back to Bentham? Explorations of perceived utility. *Quarterly Journal of Economics*, 112(2), 375–405.

Kalleberg, A.L., Rasell, E., Hudson, K., Webster, D., Reskin, B.F., Naoi, C. and Appelbaum, A. 1997. *Non-Standard Work, Substandard Jobs: Flexible Work Arrangements in the U.S.* Washington, DC: Economic Policy Institute.

Klein, G. 1998. *Sources of Power: How People Make Decisions*. Cambridge, MA: MIT Press.

Kleven, H.J., Kreiner, C.T. and Dixon, H.D. 2002. Dual labour markets and nominal rigidity. *Oxford Economic Papers*, 54(4), 561–83.

Korpi, T. 1997. Is utility related to employment status? Employment, unemployment, labour market policies and subjective wellbeing among Swedish youth. *Labour Economics*, 4, 125–47.

Krimsky, S. and Golding, D. (eds). 1992. *Social Theories of Risk*. London: Praeger.

Kunda, G., Barley, S.R. and Evans, J. 2002. Why do contractors contract? The experience of highly skilled technical professionals in a contingent labour market. *Industrial and Labour Relations Review*, 55(2), 234–61.

Levy-Garboua, L. and Montmarquette, C. 2004. Reported job satisfaction: What does it mean? *Journal of Socio-Economics*, 33, 135–51.

Malone, T.W. 2004. *The Future of Work: How the New Order of Business Will Shape Your Organization, Your Management Style, and Your Life*. Cambridge, MA: Harvard Business School Press.

Marshall, M.G. 1999. Flexible specialization, supply-side institutionalism and the nature of work systems. *Review of Social Economy*. Available at: http://www2.cddc.vt.edu/digitalfordism/fordism_materials/marshall.htm [accessed: 29 March 2006].

Martinez Lucio, M. and Stuart, M. 2005. 'Partnership' and new industrial relations in a risk society: An age of shotgun weddings and marriages of convenience? *Work, Employment and Society*, 19(4), 797.

McDonald, I.M. and Solow, R.M. 1985. Wages and employment in a segmented labour market. *Quarterly Journal of Economics*, 100(4), 1115–41.

McGovern, M. and Russell, D. 2001. *The New Brand of Expertise: How Independent Contractors, Free Agents and Interim Managers are Transforming the World of Work*. Woburn, MA: Butterworth-Heinemann.

Meyer, C. and Mukerjee, S. 2007. Investigating dual labour market theory for women. *Eastern Economic Journal*, 33(3), 301–16.

Morgan, J. 1998. Making jobs more secure: Greater employment protection makes economic sense. *New Economy*, 5(1), 44–48.

Mythen, G. 2005. Employment, individualization and insecurity: Rethinking the risk society perspective. *The Sociological Review*, 53(1), 129–49.

New Economy Task Force. 2000. *Making the New Economy Grow*. Washington, DC: Progressive Policy Institute. Available at: http://www.ppionline.org/ndol/print.cfm?contentid=1490 [accessed: 15 March 2007].

Office for National Statistics. 2009. *The Impact of the Recession on the Labour Market*. Newport, South Wales: Office for National Statistics. Available at: http://www.statistics.gov.uk/downloads/theme_labour/impact-of-recession-on-LM.pdf [accessed: 22 September 2010].

Office for National Statistics. 2010a. GDP Growth. Newport, South Wales: Office for National Statistics. Available at: http://www.statistics.gov.uk/cci/nugget_print.asp?ID=192 [accessed: 22 September 2010].

Office for National Statistics. 2010b. GDP and Unemployment. Newport, South Wales: Office for National Statistics. Available at: http://www.statistics.gov.uk/cci/nugget.asp?id=2294 [accessed: 22 September 2010].

Office for National Statistics. 2010b. Employment. Newport, South Wales: Office for National Statistics. Available at: http://www.statistics.gov.uk/cci/nugget.asp?id=12 [accessed: 22 September 2010].

Organization for Economic Co-operation and Development 1989. *Labour Market Flexibility: Trends in Enterprises*. Paris: Organization for Economic Co-operation and Development.

Osterman, P. 1984. *Internal Labour Markets*. Cambridge, MA: MIT Press.

Osterman, P. 1988. *Employment Futures: Reorganization, Dislocation and Public Policy*. New York: Oxford University Press.

Parker, R.E. 1994. *Flesh Peddlers and Warm Bodies: The Temporary Help Industry and its Workers*. New Brunswick, NJ: Rutgers University Press.

Pastoriza, D., Ariño, M.A. and Ricart, J.E. 2008. Ethical managerial behaviour as an antecedent of organizational social capital. *Journal of Business Ethics*, 8(3), 329–41.

Pfeifer, C. 2009. Fixed-term contracts and employment adjustment: An empirical test of the core–periphery hypothesis using German establishment data. *The Economic Record*, 85(268), 92–107.

Pink, D.H. 2001. *Free Agent Nation: How America's New Independent Workers are Transforming the Way We Live*. New York: Warner Business Books.

Piore, M.J. 1975. Notes for a theory of labour market segmentation, in *Labour Market Segmentation*, edited by R.C. Edwards, M. Reich and D.M. Gordon. Lexington, MA: D.C. Heath and Company.

Piore, M.J. and Sabel, C.F. 1984. *The Second Industrial Divide: Possibilities for Prosperity*. New York: Basic Books.

Polivka, A.E. 1996. A profile view of contingent workers. *Monthly Labour Review*, 119, 10–21.

Reinhold, B.B. 2001. *Free to Succeed: Designing the Life You Want in the New Free Agent Economy*. New York: Plume.

PricewaterhouseCoopers 2007. *Managing Tomorrow's People: The Future of Work to 2020*. London: PricewaterhouseCoopers. Available at: http://www.pwc.com/gx/en/managing-tomorrows-people/future-of-work/pdf/mtp-future-of-work.pdf [accessed: 24 September 2010].

PricewaterhouseCoopers 2008. *Managing Tomorrow's People. Millennials at Work: Perspectives from a New Generation*. London: PricewaterhouseCoopers. Available at: http://www.pwc.com/gx/en/managing-tomorrows-people/future-of-work/pdf/mtp-millennials-at-work.pdf [accessed: 24 September 2010].

PricewaterhouseCoopers 2009. *Managing Tomorrow's People. How the Downturn Will Change the Future of Work to 2020*. London: PricewaterhouseCoopers. Available at: http://www.pwc.com/gx/en/managing-tomorrows-people/future-of-work/pdf/mtp-how-the-downturn.pdf [accessed: 24 September 2010].

PricewaterhouseCoopers 2010. *Talent Mobility 2020. The Next Generation of International Assignments*. London: PricewaterhouseCoopers. Available at: http://www.pwc.com/gx/en/managing-tomorrows-people/future-of-work/pdf/talent-mobility-2020.pdf [accessed: 24 September 2010].

Purser, G. 2006. Waiting for Work: An Ethnography of a Day Labour Agency. *ISSC Fellows Working Papers*. Berkeley, CA: University of California Berkeley.

Available at: http://www.escholarship.org/uc/item/5vg5d05d [accessed: 15 September 2010].

Richardson, M., Stewart, P., Dandford, A., Tailby, S. and Upchurch, M. 2005. Employees' experiences of workplace partnership in the private and public sector, in *Partnership and the Management of Employment Relations*, edited by M. Stuart, and M. Martinez Lucio. Abingdon: Routledge, 210–25.

Rifkin, J. 1996. *The End of Work: the Decline of the Global Labour Force and the Dawn of the Post-Market Era*. New York: Putnam.

Robinson, A.M. and Smallman, C. 2006. The contemporary British workplace: A safer and healthier place? *Work, Employment and Society*, 20(1), 87–107.

Sabel, C.F. 1989. Flexible specialization and the re-emergence of regional economies, in *Reversing Industrial Decline? Industrial Structure and Policy in Britain and Her Competitors*, edited by P. Hirst and J. Zeitlin. Oxford: Berg.

Sennett, R. 1998. *The Corrosion of Character: the Personal Consequences of Work in the New Capitalism*. London: Norton.

Smith, N., Cebulla, A., Cox, L. and Davies, A. 2006. Risk perceptions and the presentation of self: Reflections from fieldwork on risk. *Forum: Qualitative Social Research*, 7(1), Art. 9. Available at: http://www.qualitative-research.net/fqs-texte/1-06/06-01-09-e.htm [accessed: 4 March 2007].

Smith, V. 1997. New forms of work organization. *Annual Review of Sociology*, 23, 315–39.

Smith, V. 1998. The fractured world of the temporary worker: Power, participation and fragmentation in the contemporary workplace. *Social Problems*, 45, 1–20.

Smith, V. 2001. *Crossing the Great Divide: Worker Risk and Opportunity in the New Economy*. Ithaca, NY: Cornell University Press/ILR Press.

Strangleman, T. 2007. The nostalgia for permanence at work? The end of work and its commentators. *The Sociological Review*, 55(1), 81–103.

Streeck, W. 1991. On the institutional conditions of diversified quality production, in *Beyond Keynesianism: The Socio-Economics of Production and Full Employment*, edited by E. Matzner, and W. Streeck. Aldershot: Edward Elgar.

Tilly, C. 1996. *Half a Job: Bad and Good Part-Time Jobs in a Changing Labour Market*. Philadelphia, PA: Temple University Press.

Turner, C.L. 1995. *Japanese Workers in Protest: an Ethnography of Consciousness and Experience*. Berkeley and Los Angeles, CA: University of California Press.

Twiname, L.J., Humphries, M. and Kearins, K. 2006. Flexibility on whose terms? *Journal of Organizational Change Management*, 19(3), 335–55.

Van de Ven, A.H. 2007. *Engaged Scholarship: A Guide for Organizational and Social Research*. New York: Oxford University Press.

Vogt, C.P. 2005. Maximizing human potential: Capabilities theory and the professional work environment. *Journal of Business Ethics*, 58, 111–23.

Warr, P. 1999. Wellbeing and the workplace, in *Wellbeing: The Foundations of Hedonic Psychology*, edited by D. Kahneman, E. Diener and N. Schwarz. New York: Russell Sage Foundation, 392–412.

Warr, P. and Wall, T. 1975. *Work and Wellbeing*. Harmondsworth: Penguin Books.

Whitfield, K. 2000. High-performance workplaces, training, and the distribution of skills. *Industrial Relations*, 39(1), 1–25.

Williams, K., Cutler, T., Williams, J. and Haslam, C. 1987. The end of mass production? *Economy and Society*, 16(3), 405–39.

Zhang, L. (2008). Lean production and labour controls in the Chinese automobile industry in an age of globalization. *International Labour and Working Class History*, 73(1), 24–44.

Zickar, M.J. and Carter, N.T. Reconnecting with the spirit of workplace ethnography: A historical review. *Organizational Research Methods*, 13(2), 304–19.

Zweig, M. 2000. *Working Class Majority: America's Best Kept Secret*. Ithaca, NY: Cornell University Press.

6

Aviation and Corporate Social Responsibility

Simon Bennett

Introduction

On 12 February 2009 a regional airliner operated by Colgan Air crashed into Clarence Centre, a residential neighbourhood near Buffalo, New York State. Fifty people died. Flight 3407 was operated on behalf of Continental Airlines. The primary cause of the accident was the Captain's incorrect response to a stall warning. It is likely that flight crew fatigue, induced by long-distance commutes to work and exacerbated by Heath-Robinson sleeping arrangements, played some part in the crew's degraded performance.[1] The Captain commuted to Newark from Florida. The First Officer commuted to Newark from Seattle, via Memphis. Neither crewmember had accommodation at Newark. Following the accident some commentators claimed the low wages paid by the regional airline sector had helped create a cadre of impoverished, residentially dislocated pilots who spent much of their between-duty time commuting to and from work (a recipe for acute and chronic fatigue). This chapter uses actor-network theory (ANT) and discourses on corporate social responsibility (CSR) to explore the social, economic and political dynamics of the airline industry. The work of Beck (Risk Society) and Reason (latent error/ resident pathogens) is also referenced. The analysis is holistic and inclusive. It acknowledges that aviation is subject to numerous pressures (like fuel price

---

1 According to the Federal Aviation Administration (2010), fatigue '... is characterised by a general lack of alertness and degradation in mental and physical performance'. Symptoms include 'Measurable reduction in speed and accuracy of performance; Lapses of attention and vigilance; Delayed reactions; Impaired logical reasoning and decision-making, including a reduced ability to assess risk or appreciate consequences of actions; Reduced situational awareness, and low motivation to perform optional activities' (Federal Aviation Administration 2010).

fluctuations, regulatory initiatives, competition for passengers/freight and customer preference). The chapter aims to produce safety-enhancing policy recommendations that are *practical* and *affordable*. Parallels are drawn between the Buffalo accident and a 2009 near-miss event in Europe that featured a similar fatigue-inducing long-distance commute to work (the aircraft involved belonged to a UK-registered logistics airline). It is suggested first, that for safety's sake the lifestyle issues foregrounded by the Buffalo accident must be addressed, and secondly that, as demonstrated by the near-miss, such issues are not confined to North America.

Analytical Tools

ACTOR-NETWORK THEORY

ANT provides a way of understanding complex socio-technical systems (like health care, power generation, the military-industrial complex or commercial aviation). ANT makes no value judgments. The methodology is amoral, Machiavellian even. It produces a factual record (which can be used to inform investors', company directors', politicians' or civil servants' decisions). ANT posits that socio-technical systems emerge from a process of 'heterogeneous engineering' (Law 1987). Systems consist of numerous, more-or-less-aligned human and non-human (tangible and intangible) actants (Callon and Latour 1981, Latour 2005). The closer the alignment (the more in-step the actants), the stronger and more effective the system (Stalder 1997, Miettinen 1999). ANT acknowledges reality's contiguousness (connectedness) and messiness. It acknowledges enrolled ('translated') actants' interdependence and mutual shaping (affective interaction) and expediency in human enterprise:

> The theory's aim is to describe a society of humans and non-humans as equal actors tied together into networks built ... to achieve a particular goal, for example the development of a product [like an investment bond or air service]. (Stalder 1997)

Viewed through the lens of ANT, business enterprise involves the purposeful assembly of human and non-human elements (core employees, consultants, auditors, lawyers, landlords, educators, hardware, real estate, finance markets, professional associations, rules, strategic plans, organisational cultures, etc.) to achieve some predetermined goal (for example, the manufacture of motor cars or extraction of oil from under the sea) in as efficient a manner as possible.

CORPORATE SOCIAL RESPONSIBILITY

Business enterprise is more than just a technical pursuit. It is, fundamentally, a political activity laden with value judgments. How much do you pay your shop-floor workers? How much do you pay your top executives? How much do you charge for essential commodities (like water, gas, electricity, bread and milk) for which demand is relatively inelastic? Do you only satisfy minimum legal requirements (for example, in regard to employee health and safety) or do you do more than the minimum? How broad should one's actor-network be? How closely must actants be aligned to gain maximum traction on the task or problem at hand?

During the financial crisis of 2008 the UK government mounted a successful bail-out of several of Britain's banks (Schwartz 2008, Cable 2009).[2] In 2009 bonus payments to bankers and City executives drew widespread criticism (Hiscott 2009). The Chancellor introduced a 50 per cent tax on bonuses over £25,000. In response the Conservative Party promised to tax high earners (*The Spectator* 2009). This drew a reaction from those who might be affected by the new tax (Barty 2009, Dey and Walsh 2010) and from, amongst others, the Mayor of London, who spoke out on behalf of the City of London's financial institutions.[3] Business decisions (awarding six or seven-figure bonuses, for example) can be politically highly sensitive, especially if they are, or appear to be motivated by greed. In recent years some businesspeople have made allowances for public opinion in their decision-making (Business Link 2010). As evidenced by the Brent Spar decommissioning episode, even major corporations like Shell cannot afford to ignore public opinion entirely. The boycotting of Shell garages, occupation of the Brent Spar platform, and media attention encouraged Shell to choose a more environmentally-friendly disposal option (Kirby 1998). The 2010 Gulf of Mexico oil spill again demonstrated the importance of addressing public concerns about the environmental impacts of business activity.[4] In January 2010 it appeared that the banks, unlike Shell, had no intention of responding to public opinion:

---

2   Unfortunately the bail-out augmented Britain's national debt. By 2010 the United Kingdom's national debt was four times its gross domestic product (GDP) (McRae 2010).

3   The City makes a major contribution to Britain's GDP: 'The City of London [accounts] for roughly four percent of UK GDP, out of a total financial services sector of around 9 percent of GDP' (Buiter 2008).

4   It is estimated that 4.4 million barrels of oil escaped from the British Petroleum Deepwater Horizon well-head (Pope 2010). BP is subject to reparations. The company's CEO Tony Hayward was criticised for his apparent inability to comprehend (and seeming unwillingness to atone for) the environmental and economic impacts of the spill (Simms 2010). Hayward is no longer BP's CEO. Says Andrew Simms (2010), Head of the New Economic Foundation,

> *Almost every large bank is planning to shelter employees from the 50%*
> *bonus tax. Rather than reducing bonus payouts, the banks will absorb*
> *the cost themselves. (Dey and Walsh 2010)*

In his speech to the 2010 Liberal Democrat conference Dr Vince Cable, the Business Secretary in the Conservative–Liberal Democrat coalition government, referred to bankers as 'spivs and gamblers' (Cable cited in Oborne 2010).

Today, CSR is an established business theme (Certo and Certo 2009, Business Link 2010). Davis (1973) offers this definition of CSR:

> *[Corporate social responsibility] is the firm's obligation to evaluate in its*
> *decision-making process the effects of its decisions on the external social*
> *system in a manner that will accomplish social benefits along with the*
> *traditional economic gains which the firm seeks ... social responsibility*
> *begins where the law ends. A firm is not being socially responsible if it*
> *merely complies with the minimum requirements of the law ... Social*
> *responsibility ... is a firm's acceptance of a social obligation beyond the*
> *requirements of the law.*

Bartol and Martin (1998) define CSR as 'The obligation of an organisation to seek actions that protect and improve the welfare of society along with its own interests'. The British government defines CSR as 'taking a responsible attitude [and] going beyond the minimum legal requirements'. According to the government, CSR 'can help you improve your business performance'. CSR 'reduces the risk of sudden damage to your reputation' (Business Link 2010).

Bartol and Martin (1998) identify three strains of CSR:

Perspective on CSR	Objectives
The invisible hand	To maximise profitability and shareholder value by legal means
The hand of government	To maximise social benefit by passing enlightened laws (like the Working Time Directive or Equal Pay Act)
The hand of management	To maximise profitability and shareholder value by legal means, in a way that benefits not just social elites but the whole of society

'The Deepwater Horizon disaster highlighted the economics of short-term slash-and-burn for maximum profit, and damn the consequences'.

In his seminal 1973 paper *The Case for and Against Business Assumption of Social Responsibilities*, Davis (1973) supports the notion of corporate social responsibility:

> *Assuming that the direction of business social responsibility is decided, then business institutions must move vigorously toward integrating social values into their decision-making machinery. The business which vacillates or chooses not to enter the arena of social responsibility may find that it gradually will sink into customer and public disfavour.*[5]

Certo and Certo (2009) offer the following interpretation of Davis's argument:

> *Because business is such an influential member of society ... it has the responsibility to help maintain and improve the overall welfare of society. If society already puts this responsibility on its individual members, then why should its corporate members be exempt?*

Some, like economist Milton Friedman, have argued against CSR on the basis that it may prevent corporations achieving their primary goal – the maximisation of shareholder value:

> *In a free enterprise, private property system, a corporate executive is an employee of the owners of the business. He has direct responsibility to his employers. That responsibility is to conduct the business in accordance with their desires, which generally will be to make as much money as possible while conforming to the basic rules of society ....(Friedman cited in Certo and Certo 2009)*

Under the free enterprise system a corporate executive's sole duty is 'to make as much money for their stockholders as possible' (Friedman cited in Davis 1973).

RISK SOCIETY

According to Beck (1992) we live in a *reflexive* modernity:

---

5   In coming to this view Davis may have been influenced by three major episodes in US history. First, the end of the Vietnam War, a conflict that had divided the American people. Second, his countrymen's realisation that industrial activity was polluting the environment. Third, episodes of civil unrest in urban America (the 1967 Detroit riots, for example). In Davis's mind corporate social responsibility could help rebuild civic society, promote social cohesion and reduce pollution.

> [W]hile in classical industrial society, the 'logic' of wealth production
> dominates the 'logic' of risk production, in the risk society this
> relationship is reversed ... In the welfare states of the West ... the
> struggle for one's 'daily bread' has lost its urgency ... Parallel to that
> the knowledge is spreading that the sources of wealth are 'polluted' by
> growing 'hazardous side-effects'.

Those who can afford to (the world's wealthier citizens) focus on risk reduction.
Risk management becomes a major theme of civic discourse and politics
(Beck 1992, 2009). An increasingly reflexive (questioning/sceptical/self-aware)
public seeks more control over potentially risky scientific and technological
developments. Risk Society witnesses the decline of deference. Established
authorities (scientists and church leaders, for example) are challenged.[6] A post-
traditional society emerges:

> [T]he more science and technology permeate and transform life on
> a global scale, the less this expert authority is taken as a given. In
> discourses concerning risk ... the mass media, parliaments, social
> movements, governments, philosophers, lawyers, writers, etc ... are
> winning the right to a say in decisions. (Beck 2009)

Citizens of the Risk Society look to politicians for risk management solutions
and for compensation when risks cannot be controlled (Woollacott 1998). In
the public's imagination personal security overrides personal freedom as a
legislative principle:

> Fear determines the attitude towards life. Security is displacing freedom
> and equality from the highest position on the scale of values. The result
> is a tightening of laws, a seemingly rational 'totalitarianism of defence
> against threats'. (Beck 2009)[7]

---

6   During his 2010 state visit the Pope criticised Britain's 'aggressive secularism'.

7   Following Beck's reasoning, might the Risk Society pose a threat to hard-won individual
    freedoms? Is oppression the price of security? If it is, is it a price worth paying? In Britain
    the number of closed-circuit television (CCTV) cameras has burgeoned in recent years. Until
    removed from office the Labour Party advocated a national ID card. Many motorways, dual
    carriageways and A-class roads are monitored by Automatic Number-Plate Recognition
    (ANPR) cameras, used to track vehicle movements. In September 2010 the Chief Constable
    of the West Midlands Constabulary issued a public apology (in a televised broadcast) for
    failing to consult over the installation of CCTV and ANPR cameras in two Birmingham
    neighbourhoods (the cameras were quickly covered with black plastic bags in an attempt
    to assuage public disquiet). Civil liberties campaigners claimed the cameras turned every
    resident into a suspect. Those who photograph major buildings (even if they hold a press
    card) may be questioned by police.

LATENT ERRORS AND RESIDENT PATHOGENS

The burgeoning complexity of modern technological systems (like nuclear power plants or aircraft) precludes the identification and testing of every potential failure mode. Consequently systems may harbour numerous weaknesses (latent errors/resident pathogens) which, under certain conditions, may manifest as system failures (active errors). With reference to ANT, weaknesses may reside in either human or non-human actants, or in the interfaces/interactions between actants (for example, between an airline's training syllabus, a specific flight crew and an aircraft's avionics). The fact that such weaknesses (latent errors/resident pathogens) were not identified at the design stage means that there are no procedures to resolve whatever upsets might occur. This leaves (for example) pilots or nuclear control room personnel in a difficult position: they do not know why the upset has occurred, and have no procedures to stabilise (and, hopefully, de-escalate) the crisis. They are working in the dark, as it were. The only resource they have is their knowledge and intuition.[8] For Reason (1990) concurrency – the intersection of system characteristics, outputs and human actions – plays a part in incident and accident:

> [M]ajor disasters … are rarely if ever caused by any one factor, either mechanical or human. Rather, they arise from the unforeseen and usually unforeseeable concatenation of several diverse events, each one necessary but singly insufficient … [System failure] results from a complex interaction between latent failures [resident pathogens] and a variety of local triggering events.

Case Study 1

While the immediate cause of the Buffalo crash was pilot error, there were several proximate or contributory causes, including deficiencies in stall-recovery training and, on balance of probabilities, crew fatigue. The Captain, who was pilot-in-command for the approach into Buffalo, was on the third day of a duty that began at 05:45 on 10 February. The previous day he had commuted to Newark (EWR) from Tampa, Florida, arriving at 20:05 to spend the night in

---

8   This is an argument for the retention of human operators in complex systems (like aircraft, high-speed trains or oil production platforms). Automation may be motivated more by economic than safety concerns. Brookes (2002) says: '[A]utomation must never reach the stage where the human is pushed too far out of the loop. The difference between safety and catastrophe is usually an alert someone who … is able to respond flexibly to whatever fate throws at them'.

Figure 6.1   Fifty died in the Flight 3407 accident
Source: Photo by David F. Sherman, copyright Bee Group Newspapers.

Colgan Air's Crew Room. It has been suggested that the Captain's salary was sufficient for him to rent accommodation or take a hotel room in New Jersey (Turner 2010). According to the NTSB (2010) '[Captain Renslow] was trying to get around having to pay for a crash pad by bidding trips that had overnights or commutable ends'. In short, Renslow looked for duties that a) provided him with as much down-route hotel accommodation (HOTAC) as possible, and b) finished at a time that allowed him to immediately catch a flight back to Florida.

Fifteen per cent (20/137) of Colgan's Newark-based pilots lived 400–1,000 miles from base, while 21 per cent (29/137) lived more than 1,000 miles from base (Rosenker 2009). Rationalisation has helped create a pilot diaspora (McGreal 2010). Early in 2009 Erlanger-based regional carrier Comair's parent company Delta 'shifted the majority of Comair's operations to the Northeast ... forcing many locally-based pilots to commute north to report for work' (*Bloomberg News and The Enquirer* 2009). McGreal (2010) says: '[P]ilots [are] reluctant to uproot their families, pull children out of school and sell houses only to be moved again in a year'.

On 10 February Renslow flew three sectors (EWR-YYZ-EWR-BUF) before checking in to his downroute hotel (HOTAC) in Buffalo. On 11 February he reported at 06:15 to fly three sectors (BUF-EWR-RDU-EWR). With brakes on at 15:44 he faced a long second night resting/sleeping in the Crew Room. He logged into his airline's crew scheduling system at 03:10 and 07:26. He reported for duty at 13:30, but, due to bad weather, had to wait until after 21:00 to fly his first sector of the day. According to the NTSB (2010), at the time of the accident Renslow 'had a cumulative sleep debt of between 6 and 12 hours'. This may have affected his performance.

Asked about pilots' Heath-Robinson sleeping arrangements Harry Mitchel, Colgan's Vice President for Flight Operations, told the National Transportation Safety Board (NTSB): 'It's not quality rest. There's a lot of activity in our Crew Rooms' (Mitchell cited in Wald 2009). To retain his Federal Aviation Authority (FAA) Medical Certificate the Captain was required to have a blood test every six months. Captain Renslow took Gemfibrozil (for cholesterol), Diltiazem (for hypertension/angina) and the diuretic Hydrochlorot (National Transportation Safety Board 2009a). He was 47 years old.

Renslow's First Officer, Rebecca Shaw, commuted to Newark from Seattle, a journey that took seven and a half hours. Two flights were involved, each made as a jump-seat passenger on FedEx freighters. Shaw managed to sleep on the flights. On arriving at Newark she slept in the Crew Room – something Colgan Air had forbidden its pilots to do (Rosenker 2009). Between the time she woke in Seattle on 11 February and the time she reported for duty in Newark, Shaw managed to get around six hours' sleep (either on jump seats or on Crew Room loungers). The elapsed time between her waking in Seattle and the impact time was in excess of 33 hours. Shaw's movements are recorded below:

	11 Feb			12 Feb	
Shaw woke	09:00–10:00	PST			
Depart Seattle (FedEx Flight 1223)	20:00	PST			
	(23:30)	PST	**Arrive Memphis**	02:30	EST
	(01:20)	PST	**Depart Memphis (FedEx Flight 1514)**	04:20	EST
	(03:30)	PST	**Arrive Newark**	06:30	EST
	(10:30)	PST	**Report**	13:30	EST
	(18:18)	PST	**Cleared for take-off**	21:18	EST
	(19:17)	PST	**Impact**	22:17	EST

PST = Pacific Standard Time; EST = Eastern Standard Time

Unlike Renslow, Shaw commuted to work on the day she reported. At the time of the accident Colgan Air did not discourage this practice. Colgan's *Flight Crewmember Policy Handbook* originally stated: 'Flight Crewmembers should not attempt to commute to their base on the same day they are scheduled to work'. This advice did not appear in Colgan's March 2008 handbook – the edition in use at the time of the Buffalo accident (NTSB 2010). This raises two questions. First was Colgan Air's safety culture improving or decaying? Second, did the underpinning philosophy of Colgan's *Flight Crewmember Policy Handbook* shape its pilots' commuting behaviour? Human behaviour is shaped both by cultural norms (which are tacit/undocumented) and by rules and regulations (which are explicit/documented).

During taxi-out Shaw told Renslow she felt unwell. The conversation was captured on the aircraft's cockpit voice recorder (CVR) (Lowy 2009b):

Shaw	I'm ready to be in the hotel room ... This is one of those times that if I felt like this when I was at home there's no way I would have come all the way out here. But now that I'm out here ...
Renslow	You might as well
Shaw	I mean, if I call in sick now, I've got to put myself in a hotel room until I feel better ... We'll see how it feels flying. If the pressure is just too much I ... could always call in [sick] tomorrow. At least I'm in a hotel on the company's buck, but we'll see. I'm pretty tough.

Shaw's reference to the cost of HOTAC suggests she had money worries. Colgan paid degree-holding Shaw less than $24,000 per annum. When she started flying for Colgan early in 2008 she was paid less than $17,000. The following wage levels obtained in May 2008 (United States Bureau of Labour Statistics 2008):

Occupation	Mean annual wage $US
Dishwasher	17,700
Short order cook	20,200
Baggage porter/bellhop	23,200
Filing clerk	25,300
Construction labourer	32,200
Bus driver	35,700
Carpet installer	41,300
Carpenter	42,900
Brickmason	47,700

In the United States it is common practice for pilots to commute long distances. The practice (which Colgan constructed as a 'freedom' (Colgan Air 2009)) is apparently supported by both employers *and* pilots: '[T]he airlines pay burger-flipping wages to pilots, who commute and are fatigued. The pilots don't want any restrictions on their commuting ... so their unions don't question airline schedules ...' (Evans 2009). Pilots' ability to 'dead-head' on commercial aircraft underwrites the commuting culture. Pilots who are starting their careers (like Shaw) or who find themselves displaced (perhaps because they were made redundant or furloughed) may choose to commute rather than buy or rent accommodation near their base. Long-distance commuting is a by-product of the aviation industry's volatility (McGreal 2010) and, since deregulation in the late 1970s, tendency to bear down on wage costs (Koehler, Esquivias and Varadarajan 2009, Massachusetts Institute of Technology 2011). Many of those who fly for regional carriers like Colgan Air earn between $20,000 and $30,000 (Bennett 2009, McGreal 2010, Ray 2010). Jeppesen's Senior Training Manager comments: 'You're looking at low pay and a lifestyle dictated by that pay level' (Wright cited in Ray 2010). Because the majors (who can pay well) have not recruited for about ten years, new entrant pilots are corralled into the regional, freight or parcel express sectors where salaries tend to be much lower (Ray 2010).[9] NTSB Board Member Deborah Hersman's comments about the interplay between industry flux, salary levels and homemaking are germane:

> *Flight crew commuting is particularly challenging. A regional flight crew's home base changes often, and to offset the disruption of frequent relocations, pilots may commute from a home location. The Colgan Air pilots were commuting pilots. Both pilots were based in EWR but the captain lived in Florida and the first officer in Seattle. During the previous 14 months, the first officer lived in Phoenix (when hired by the company), then expected to be based in Houston before being sent to Norfolk, Virginia, and then at the time of the accident, was based in Newark, New Jersey but lived in Seattle, Washington. Flight crew salaries are also problematic. It is financially challenging for pilots, whether earning $60,000 or $16,000, to regularly relocate their families or hold down multiple residences. When the FAA convened the fatigue ARC [Aviation Rulemaking Committee] in the summer of 2009, they took commuting off the table, and neither Colgan, nor ALPA [Airline*

---

9   Air Transport World journalist Lisa Ray (2010) observes: '[T]he average ALPA captain at a U.S. legacy airline is 52 and earns about $155,000 annually after 21 years of service, while an average first officer is 45 with 12 years of service and makes about $105,000 ... an average first officer at a U.S. regional with one year of service earns only around $20,567. A ten-year regional captain can expect to earn $70,000'.

> *Pilots Association] addressed the issue of commuting in their accident*
> *submission documents, even though 70% of the pilots based in Newark*
> *commute and 20% commute from over 1,000 miles away. I recognize*
> *that an objective analysis of commuting will be a difficult, and perhaps,*
> *uncomfortable discussion. But we should not be afraid to confront this*
> *issue in the context of understanding fatigue and its effect on pilot*
> *performance. (Hersman cited in National Transportation Safety Board*
> *2010)*

Colgan expected its pilots to rent or buy accommodation close to base. Given that accommodation costs in airport catchment areas are often much higher than elsewhere (Evans 2009, Federal Aviation Administration 2010), it could be argued that Colgan's wage levels made it difficult – if not impossible – for many of its pilots to abide by its accommodation rules. In September 2009 the FAA, reacting to the circumstances of the accident, issued a Safety Alert for Operators (SAFO) (Federal Aviation Administration 2009) that addressed the issue of crew rest/sleeping facilities:

> *Effective sleep opportunities are a critical countermeasure to fatigue ...*
> *[Airlines] should consider providing crew rest facilities that have rooms*
> *away from the general traffic for quiet, comfortable and uninterrupted*
> *sleep ...*[10]

In its final report on the accident, the NTSB (2010) suggested that airlines take a greater interest in pilots' travel-to-work and accommodation arrangements:[11]

> *Companies can take actions to help mitigate fatigue in commuting pilots.*
> *Such actions include providing rest facilities, providing assistance to*
> *pilots in identifying affordable accommodations, planning flight schedules*
> *that support commuting without extended times of wakefulness, and*
> *considering ways to evaluate and account for the effect of commuting on*
> *subsequent duty periods ... Operators have a fundamental responsibility*
> *to support their pilots' efforts to mitigate fatigue.*

The Chief Pilot at Colgan Air's Newark hub said he did not know how many of the airline's Newark-based pilots commuted to work (NTSB 2010). Without

---

10 In September 2010 the FAA (2010) published its Notice of Proposed Rulemaking on flight and duty times with the aim of having a new rule in place by August 2011.

11 In 2010 NASA proposed a study that would examine the relationship between commuting and fatigue (for both flight and cabin crew) (Croft 2010).

Figure 6.2   A corner of Clarence Centre was razed
Source: Photograph by David F. Sherman, copyright Bee Group Newspapers.

this information there was little chance he could effectively 'support pilots' efforts to mitigate fatigue'.

Many large airlines purchase services from smaller, specialist airlines. Three US majors – US Airways, United Airlines and Continental – purchased services from Colgan (Crook 2009). By lowering its cost base Colgan made itself more attractive to client airlines. Majors like Continental drive costs down *by proxy* (Bennett 2009).[12] Colgan was a cost-effective actant within Continental's actor-network.

Case Study 2

In the autumn of 2009 a fatigued freighter crew commenced a take-off run without setting their Boeing aircraft's flaps to the required 15°. The ensuing Configuration Warning (CW) alarm caused the Captain (pilot-in-command

---

12 Sub-contracting to reduce costs is commonplace: at the time of its destruction in 1988 only 17 per cent of those on-board the Piper Alpha oil production platform were Occidental employees. Occidental was criticised for its lax safety culture (Cullen 1990).

(PIC)) to bring the aircraft to a stop. The Captain described the emergency stop in his Safety Report: 'The aircraft started to move slowly until the engines spooled up to take-off power, when the configuration warning started to sound. I stopped the aircraft gently, and we realised only then that the flaps had not been set to take-off configuration'.

It is rare for a flight crew to forget to set take-off flap. Operators provide pilots with standard operating procedures (SOPs) and checklists to ensure that aircraft are configured for take-off. Pilots are trained to monitor each other's actions and are encouraged to question decisions/actions they consider unsafe (Krause 1996). On this occasion the PIC requested 'flaps 15° for take-off' but omitted to check that his First Officer (pilot monitoring (PM)) had executed his instruction. Shortly afterwards the Captain asked his First Officer to read out the Before Take-Off Checklist. The Captain missed the configuration error a second time: 'When the [First Officer's] challenge was "Flaps", I have a vivid recollection of looking at the flaps indicator[13] and "seeing" flaps set at 15°'. Because the Captain was feeling tired he subsequently performed his own non-standard before take-off check. The error was missed a third time: 'As a rule, and especially when I feel tired, I supplement our company procedures by doing a FATS [flaps, airbrakes, trim, speed] check on line-up. Once again my brain sees what it wants to see rather than facts, and I managed to oversee [overlook] once again the wrong flaps setting'. Having failed to select take-off flap the Captain failed to query why he could not select the EPR (thrust-control) button: 'When cleared for take-off I depressed EPR to activate the autothrottle system into take-off, go-around mode. It did not engage. At that point I remember very well that I decided to take-off anyway, with the vague feeling that I should not'. He continued the take-off.

The Captain's five-day duty commenced on a Sunday evening. Getting to work involved a 12-hour train journey across central Europe. That Sunday he rose at 05:30 (having had about six hours' sleep). He managed to get about one hour's sleep during the train journey. He made two changes before arriving at his base behind schedule. He rushed to his aircraft to fly one sector. That evening he felt 'whacked'. He slept from 22:00 to 08:00 on Monday morning. On Monday evening/Tuesday morning he flew three sectors. On Tuesday he slept from 09:00 to about 14:00. He filed a Fatigue Report with his airline. He wrote: 'I flew the last leg [sector]. On approach the state of fatigue was such that

---

13   The analogue flaps indicator on the Boeing 757 is located on the front panel to the right of the centre binnacle. It consists of a small circular dial marked 1,5,15,20,25,30 for the aircraft's six possible flap settings. A needle indicates the flap setting.

I had great difficulties to complete routine tasks …'. This level of impairment suggests the Captain was suffering from fatigue (Caldwell and Caldwell 2003, Federal Aviation Administration 2010). At home the Captain got eight hours' sleep each night. Down-route he got less than six.[14]

On Tuesday evening/Wednesday morning the Captain flew three sectors. On Wednesday he slept from 05:30 to 11:00. On Wednesday night/Thursday morning he flew three sectors. On Thursday he slept from 06:30 to 11:30. The CW occurred on the first sector of that Thursday evening's duty. Following the event the Captain filed a second Fatigue Report. He wrote:

> *After my first night on duty … I never managed to recuperate properly. In fact, I have been operating every night in an abnormal state of fatigue … I advised my First Officer that I felt abnormally tired, and I decided to take extra vigilance [sic] … Despite all that, the wrong flap setting went unnoticed and managed to pass through four different filters …*

The Captain was under psychological pressure at the time, having separated from his wife.

In such situations the other pilot is expected to trap any errors made by her/his colleague (Krause 1996). In this case the First Officer was both fatigued and psychologically distracted. On Tuesday night/Wednesday morning the First Officer flew two sectors. On Wednesday he got around five and half hours' sleep. On Wednesday night/Thursday morning he flew three sectors. On Thursday he got five hours' sleep. Prior to reporting for duty he received a telephone call from his wife who told him that both she and their son had a fever. Following the CW event the First Officer filed a Fatigue Report. He wrote: 'Now we both (the Captain and me) realise the lack of sleep'. Like his Captain, the First Officer was also under psychological pressure – his wife and child were sick.

The Captain's final comment on the event reveals how shocked he was at his oversight: 'I would never have thought that something like this could have happened to me … I have been flying for nearly 25 years and I have an

---

14 In their survey of fatigue levels amongst night-freight pilots Gander *et al.* (1998) found that 'During daytime layovers, individual sleep episodes were about three hours shorter than when crewmembers were able to sleep at night … Crewmembers were three times more likely to report multiple sleep episodes (including naps) on duty days than on non-duty days … Nevertheless, these additional sleep episodes were insufficient to prevent most crewmembers accumulating sleep debt across trip days'.

absolutely clean safety record. I am a strong believer in CRM and Company SOP observance. I am absolutely amazed to find out that, despite strict observance of SOPs, application of CRM to the best of my abilities and basic airmanship, such a thing still manages to slip through our attention when our mind has reached a certain stage of fatigue'.

Active schema may have contributed to the oversight. Schema are effort-efficient cognitive shortcuts that speed mental processing. Stereotypes and archetypes are forms of schema. Schema-driven reasoning has a predictive/projective quality. Schema-driven expectations may cause subjects to recall objects that were not, in fact, present (Brewer and Treyens 1981) or recall conversations that did not, in fact, take place. Reason (1990) observes: 'These stored routines are shaped by personal history and reflect the recurring patterns of past experience'. Because the Captain and First Officer nearly always remembered to select take-off flap, there was no reason to think they had forgotten on this occasion. Habituation generates a mental model of normal instrument settings. This may cause 'aberrant' settings to be overlooked or dismissed (Klein 2007). In this situation a state of mindlessness exists (Langer 1989). Subjects are unreflective. They do not interrogate their motivations/actions. (The antithesis – mindfulness – is predicated on a love of detail, intellectual curiosity, reflexivity, alertness, sensitivity to perturbations, deference to expertise (regardless of where it resides) and a commitment to resilience (Weick and Sutcliffe 2001, Hopkins 2002)).

Understanding Aviation Industry Behaviour

Airlines operate within a regulatory context, the fundamental principles of which are laid down by the International Civil Aviation Organisation, a United Nations body. Individual nations or regions (like the European Union) have their own national or regional regulatory authorities. Regulatory authorities may be responsible for both the safety and economic performance of their respective airline industries (Sochor 1991, International Civil Aviation Organisation 2003). Conflicts of interest may arise in 'dual responsibility' authorities (Bennett 2008). Following the Flight 3407 accident a Federal Aviation Administration inspector spoke about, in Lowy's (2009a) words, 'a cosy relationship between the agency [FAA] and the airline [Colgan]'. The inspector, a former pilot, had 40 years' experience in the industry. The FAA had relieved him of his Colgan brief after complaints from the airline.

Behaviour is shaped by ownership patterns. State-owned airlines may be subject to political pressures (Dienel and Lyth 1998, Bennett 2006), while private concerns must respond to market and shareholder pressures. The process of deregulating the airline industry, begun in the US in the late 1970s, continues (reflecting a broader laissez-faire zeitgeist (Cable 2009)). Behaviour is also influenced by the insurance industry (Beck 1992), organisational culture, trades unions and societal expectations (zeitgeist). The combined effect of investor pressure, the 'bonus culture' (where managers' salaries reflect financial performance), deregulation, easier market entry for start-ups and passenger demand for ever-lower fares has encouraged managers to cut costs (Koehler, Esquivias and Varadarajan 2009, Massachusetts Institute of Technology 2011). Like vacuum cleaner, refrigerator or automobile prices, airline ticket prices have dramatically fallen in real terms. Speaking in August 2009, William Swelbar (cited in Freeman 2009a), a researcher at the International Centre for Air Transportation at the Massachusetts Institute of Technology said:

> When adjusted for inflation over the last 30 years, fares are down some 50-plus percent. And that just does not make for a sustainable business model. It doesn't make a model that allows them [the airlines] to compensate their people well, like they have in the past.

Addressing the potential consequences for airline viability and safety of falling ticket prices, the vice president of the United States Air Line Pilots Association (USALPA) said: 'The public needs to ask the question: is it worth it to always look at prices as the driving factor?' (Rice cited in McGreal 2010).[15] The downward pressure on salaries (including pilots' salaries) and erosion of benefits (Sullenberger 2009, International Transport Workers' Federation 2009, Harvey and Turnbull 2009) has tended to alienate employees from managements (British Broadcasting Corporation News 2009). According to the Centre for Asia-Pacific Aviation (2009), 2010 would be a year of industrial confrontation.

---

15 Despite the intense focus on costs (Massachusetts Institute of Technology 2011), the aviation industry's safety record shows a steady improvement. Seen in a broader context flying is a remarkably safe form of transport: 'It is twenty times safer to get airborne in a commercial airliner than to drive to the airport in the first place' (Brookes 2002). The reason flying is perceived as dangerous is because for many it represents a dread risk. Having to place one's life in the hands of others (in this case, pilots and controllers) creates feelings of dread. Media coverage of airliner crashes (which often fails to mention the industry's good safety record) reinforces the dread. Although driving is significantly more dangerous than flying, the fact that one is in control mitigates feelings of dread. Motorists tend to overlook the fact that up to 3,000 persons die on Britain's roads each year.

With reference to ANT, within-company actants' alignment has degraded, weakening airlines. In December 2009 a planned 12-day strike by British Airways (BA) cabin crew was avoided only after the company sought a judicial review. BA, whose pension deficit outstrips its market value, plans to shed thousands of jobs (Fortson 2010). By the end of 2010 industrial relations at BA were in a poor state. Industrial action by cabin crew has cost BA about £150m (British Broadcasting Corporation News 2011).[16]

In May 2009 first and second-year First Officers at Colgan Air's parent company, Pinnacle Airlines Inc., earned less than $22,000 a year. Speaking in May 2009, Pinnacle Airlines Captain Amy Kotzer (cited in Risher 2009) claimed: 'Almost half our pilots earn less than $30,000 a year [less than, for example, a construction labourer]'. In February 2009 Captain Chesley Sullenberger, the pilot who safely ditched his stricken passenger aircraft in the Hudson River and was subsequently lionised by the public and media, addressed the US House of Representatives on the subject of flight crew terms and conditions:

> *Americans have been experiencing huge economic difficulties in recent months — but airline employees have been experiencing those challenges, and more, for the last 8 years! We have been hit by an economic tsunami. September 11, bankruptcies, fluctuating fuel prices, mergers, loss of pensions and revolving door management teams who have used airline employees as an ATM have left the people who work for airlines in the United States with extreme economic difficulties ... It is my personal experience that my decision to remain in the profession I love has come at a great financial cost to me and my family. My pay has been cut 40%, my pension, like most airline pensions, has been terminated ... (Sullenberger 2009)*

Providing for Pilots' Needs

In certain respects the breadth, depth and character of an air carrier's actor-network is a matter of choice. Some airlines, perhaps conscious of the impact of geographical displacement and wage reductions, provide sleeping facilities for between, or off-duty pilots. The airlines featured in the case studies employed pilots who lived significant distances from base. The freight airline had a company flat, and provided bunk spaces at its major bases. While a small charge was levied for use of the flat, the bunk accommodation was free (although there

---

16 In March 2011 BA cabin crew voted for further industrial action (Pank 2011).

were usually at least two beds to each cubicle). Colgan Air provided no such accommodation at its bases. Managers' concern for pilots' welfare was minimal – the least they could legally get away with (Bennett 2009). Colgan's pilots were expected to rent or buy accommodation, or 'double-up' with colleagues – regardless of their ability to pay for accommodation in the vicinity of the base. To ensure that pilots met their accommodation responsibilities they were banned from sleeping on company premises – despite the fact that the airline provided lounges with recliners and couches.

With regard to accommodation, the freight airline defined its responsibilities more liberally. Efforts were made to meet pilots' welfare needs. The airline's actor-network included the property market (which supplied the company flat) and on-site sleeping accommodation.

Both airlines operated services for other entities. Colgan provided wet-lease services to Continental, United and US Airways. The freight airline provided wet-lease services to a German-based forwarder. Although both airlines paid close attention to running costs, they treated their pilots differently, a reflection, perhaps, of different interpretations of corporate social responsibility. The low salaries paid by Colgan and treatment of its commute-to-work pilots suggest the airline subscribed to Bartol and Martin's (1998) 'Invisible hand' model of CSR. Continental's tolerance of Colgan's practices suggests that it, too, subscribed to the same model. Colgan and Continental pared down costs to maximise shareholder value – common practice in today's de-regulated aviation industry (Koehler, Esquivias and Varadarajan 2009, Massachusetts Institute of Technology 2011).

While the freight airline did more for its pilots than Colgan, it could not be certain that crewmembers would avail themselves of the facilities. Despite his employer's provision of rest/sleeping facilities the Captain involved in the near-miss thought it acceptable to go straight to work after making a 12-hour train journey. Even socially responsible employers cannot legislate for employees' bad judgment. The seeds of error were probably sown during that long first day.

Addressing the Commute-to-Work Issue

Colgan Air's internal investigation into the 12 February accident identified the crew's 'loss of situational awareness and failure to follow Colgan Air training and procedures' as the primary cause of the accident. Colgan identified four contributory or secondary factors. Two pertained to the design of the aircraft. Two pertained to *in situ* flight crew failings (Crook 2009). Neither failing was attributed – either wholly or in part – to flight crew fatigue. Later in his investigation report, the airline's Director of Safety, Mike Crook (2009), stated:

> *It appears that FO Shaw did not have time for adequate sleep in the 24 hours preceding the accident … Based on comments of FO Shaw on the CVR … it is apparent that her health deteriorated at some point prior to the flight … Despite these facts, it is unclear if fatigue or illness played a role in the accident.*

Crook noted of Renslow's sleep history:

> *[H]is primary rest period for the night of February 11 was likely less than five hours. Despite having ample time off duty in the three days preceding the accident, it appears Captain Renslow averaged less than six hours of overnight sleep in that time period. Accordingly, it is possible he was fatigued during the accident flight.*

On balance of probabilities, and despite Crook's analysis, it is likely that fatigue (and in Shaw's case, illness) played a part in the accident, specifically in the crew's loss of situational awareness, Captain Renslow's incorrect response to the stick-shaker and Shaw's failure to query/correct her Captain's decisions and actions. According to one expert witness Renslow's actions were consistent with him believing it was the aircraft's tail that was in a stall. He may have become fixated (Dismukes cited in NTSB 2009c). Fixation is a by-product of fatigue (Dismukes cited in NTSB 2009c). In 1999 the National Aeronautics and Space Administration (NASA) surveyed pilots at 26 regional carriers. Eight out of ten pilots admitted to 'nodding off' during flight.

The second case study suggests that fatigue affects performance. Both the Captain and First Officer believed themselves to be fatigued. The Captain had undertaken a 12-hour commute to work before reporting for duty. The Captain and First Officer had slept poorly when down-route. The First Officer was under psychological pressure. The Captain's analytical frame caused him to either overlook or ignore contra-indications (like his inability to select EPR). The First Officer's degraded performance caused him to either overlook or dismiss his Captain's errors. Both pilots seemed to be fixated on getting their aircraft into the air. Both seemed to be in a state of *mindlessness* (Langer 1989). Effective crew resource management requires the attentive monitoring and cross-checking of colleagues' reasoning, decisions and actions (Krause 1996). Effective CRM is predicated on mindfulness.

The Crash-pad Actant – Symbol of Industry Dysfunction?

One by-product of pilot dislocation is the crash-pad (perhaps 'flop-house' would be a more apt description of pilots' temporary accommodation). There are between 500 and 1,000 crash-pad houses in the United States. Renting a bed (often a space in a bunk-bed) costs from $200 per month. Beds can also be rented by the night. Sometimes the bed turns out to be an air mattress. *The Washington Post's* Sholnn Freeman (2009a) came across a suburban house that contained 30 bed spaces, 16 of which were in the basement. Curtains provided a measure of privacy.[17] Crash-pads can be depressing places: 'The interior is nondescript. The faded carpets, brownish wallpaper and secondhand furniture give rooms the feel of a low-budget motel'. According to Freeman 'A few [pilots] … complain they make so little money that they have to make crash pads their primary homes'. The pilots said the crash-pad network (the website crashpads.com has 10,000 subscribers) symbolised the dysfunctionality of the US aviation industry (Freeman 2009a). In a July 2009 article on crash-pads the *New York Post's* Sarah Ryley (2009) wrote:

> *They sleep eight to a room in cramped apartments … But it's just a day in the life for hundreds of pilots who fly out of city airports and spend their nights in dingy Queens [New York] crash pads crammed with up to 20 beds. Crew members for several major airlines told The Post they often get no more than four hours of sleep before their shifts. Some rented apartments in Kew Gardens and Jackson Heights have rooms stuffed with as many as four bunk beds. Cash-strapped pilots, mechanics and flight attendants pay as little as $125 to $260 per month to stay there.*

Most of the airlines contacted by Ryley declined to comment on her findings. The exception was American Airlines, who said: 'We are not aware of any conditions such as you describe' (Raynolds cited in Ryley 2009).

Unfortunately, according to National Transportation Safety Board Member Deborah Hersman, there is a reluctance at *every level* of the aviation industry

---

17 The FAA (2010) defines suitable accommodation as '… a temperature-controlled facility with sound mitigation that provides a crewmember with the ability to sleep in a bed and to control light'.

to confront the issue of long-distance commutes to work: 'I recognize that an objective analysis of commuting will be a difficult, and perhaps, uncomfortable discussion. But we should not be afraid to confront this issue in the context of understanding fatigue and its effect on pilot performance' (Hersman cited in National Transportation Safety Board 2010). Airlines reserve the right to hire pilots from where they choose. Pilots reserve the right to live where they like. Choice is constructed as an inalienable right. Clearly while the airlines, including Colgan Air, bear no *direct* responsibility for the crash-pad phenomenon, it could be argued that low industry salaries (Evans 2009) create a market for low-rent, low-quality, high-density accommodation (that is sometimes insanitary). Put another way, the crash-pad is the property-market's response to a set of social and economic conditions (dislocation and impoverishment) created by US airlines. *All* airlines are complicit.

Actor-networks may share the same actants, albeit under different conditions. Crash-pads are formally aligned with landlords' actor-networks and informally aligned with airlines' actor-networks. They are the purposeful creation of the property market from which landlords *and* airline managements benefit. Landlords (or, perhaps, 'slumlords') benefit from inflated rental income for minimal outlay. Airline managements benefit because crash-pads meet itinerant employees' accommodation needs – albeit poorly. Shortly after *Post* reporter Freeman's article appeared, officials from the local authority (Loudoun County) visited the 30 bed-space property in the company of Sheriff's Deputies (Freeman 2009c). While it may be correct in law to investigate the landlord for code violations, is there not a moral case for investigating those whose decisions create the demand for such accommodation, namely airline managements? Are not crash-pads airline managements' *proxy* actants?

Long-distance Commuting – What Can Be Done?

The evidence presented here suggests that long-distance commutes to work, Heath-Robinson sleeping arrangements and pilots' own poor decisions (see Case Study 2) constitute latent errors or resident pathogens (Reason 1990) in the national airspace systems of both the United States and the European Union. It suggests that 'corporate social responsibility' means different things to different airlines. In terms of welfare provision Colgan Air did the least it could legally get away with for its pilots. Colgan's Friedmanesque management style stands in sharp relief to the freight airline's more philanthropic approach.

The freight airline did more than the minimum, providing sleeping accommodation both on and off-site. In the UK there is a long tradition of employer-subsidised housing (Titus Salt built Saltaire, the Cadbury brothers built Bournville and British Petroleum built the village of Llandarcy adjacent to Britain's first oil refinery). The freight airline provided a company flat. Colgan Air warned its poorly paid pilots that if they were found sleeping on company premises they might be dismissed. Colgan showed little concern for the plight of its low-paid, dislocated pilot-workforce. It appears that Colgan's managers had no interest in the social and economic outcomes of their decisions (like hiring dislocated pilots on low wages, banning them from sleeping in Colgan's crew rooms and tolerating the exploitative and insanitary crash-pad rental sector). As far as Colgan's managers were concerned, *where* a pilot resided was her/his lifestyle choice (National Transportation Safety Board 2009b, Colgan Air 2009). Bennett (2009) states: 'Colgan Air's managers made no attempt to understand the wider socio-economic environment. They divorced themselves from the lived reality of the flight crew lifestyle. They lived in a bubble'. Politicians, too, failed to understand that industry volatility, low wages and indebtedness (the product of university and/or flight training fees) meant many pilots had little choice but to commute long distances to work. Republican Tom Petri's comments were typical. Petri (2010) framed commuting as a lifestyle choice:

> Clearly, commuting is a part of the lifestyle choice for airline pilots ... [I]f we agree that irresponsible commuting is a causal factor in fatigue, then the practice of commuting deserves a look.

It could be argued that the irresponsible parties are employers who pay low salaries to indebted pilots who are then posted to bases hundreds or thousands of miles from home. The FAA, in its post-Colgan NPRM, acknowledged the impact of economic and other factors on pilot behaviour:

> Commuting is common in the airline industry, in part because of lifestyle choices ... but also because of economic reasons associated with ... frequent changes in the flightcrew member's home base, and low pay and regular furloughs [lay-offs] by some carriers that may require a pilot to live someplace with a relatively low cost of living. (Federal Aviation Administration 2010)

Some airlines endeavour to accommodate their pilots' needs, others do not. Airline human resource departments' responses to the global financial crisis that commenced in 2008 showed great variation: '[Responses] ranged from the immediate and unilateral to the considered and consultative' (Harvey and Turnbull 2009). Airlines interpret 'social responsibility' differently.

While the freight airline did more to meet its pilots' welfare needs, it too hired pilots who lived many hundreds of miles from base. Because some pilots saw nothing wrong with going straight to work after a long commute, this policy helped erode safety margins. If pilots won't use the sleeping facilities provided the only solution is to hire locally. Of course, while this may be possible during a downturn (when there are more pilots than jobs) it is more difficult when times are good and pilots are a scarce commodity.

The Flight 3407 accident drew significant comment (supporting Beck's thesis that citizens are more risk-conscious). The following comments were posted on the internet (Risher 2009):

> *[Regional] pilots are extremely professional and are no less responsible for their passengers' well-being than those of major airlines. They deserve decent work rules and a decent living wage.*

> *[T]he REAL responsibility lies squarely with the management, no matter how much they try to assign all the blame to the pilots.*

> *That level of salary is unacceptable for someone who has that much responsibility.*

John Boccieri (cited in Zremski and Schulman 2009), a member of the House of Representatives Transportation Committee and a pilot in the United States Air Force Reserve said: 'The committee I serve on has to seriously address what's happening to commercial aviation in this country'. A relative of one of the victims said: 'There seems to be truly an indifference to commuting, to pay, to fatigue' (Mellett cited in Freeman 2009b). *Buffalo News* reporters Zremski and Schulman (2009) noted: 'Regional airlines now run nearly half of the nation's commercial flights. But those airlines ... have been responsible for all of the nation's multiple-fatality commercial plane crashes since 2002'. National Transportation Safety Board member Kathryn O. Higgins (cited in Wald 2009) commented: 'When you put together the commuting patterns, the pay levels, the fact that your crew rooms that aren't supposed to be used, are being used, I think it's a recipe for an accident'. *The New York Times* (2009) addressed the Buffalo accident in an editorial:

> *Reports have emerged of poorly paid commuter pilots who hopscotch across the country to work and sleep wherever they can. They sometimes sack out in lounge chairs in airports or on the floors of planes or even in their cars ... [C]ommuter pilots are flying too much, sleeping too little and placing passengers at risk.*

An airline employee wrote the following on the web site About.com (2009):

> *What many people don't realise is how many airline employees do commute. Not just pilots, but flight attendants, airport agents ... I certainly know what it's like to ... sleep on a couch in an employee lounge and live off caffeine ... commuting cross-country is not something that is without impact on one's body.*

While such reactions are understandable, it is important to remember that aviation's *modus operandi* is partly a reflection of the travelling public's desire for cheap travel. Yes, people are more risk-aware. But they are not *so* aware that they are willing to consider the consequences of cost-focused consumerism. When margins are squeezed (for example by rising fuel prices, higher insurance costs, higher landing charges or the elastic (price-sensitive) demand for airline seats) executives have no choice but to cut overheads. Salaries fall. Morale deteriorates. Sullenberger's (2009) observations are pertinent:

> *[T]he terms of [pilots'] employment have changed dramatically ... leading to an untenable financial situation for pilots and their families. When my company [US Airways] offered pilots who had been laid off the chance to return to work, 60% refused ... I am worried that the airline piloting profession will not be able to continue to attract the best and the brightest. The current experience and skills of our country's professional airline pilots come from investments made years ago when we were able to attract the ambitious, talented people who now frequently seek lucrative professional careers. That past investment was an indispensible element in our commercial aviation infrastructure, vital to safe air travel and our country's economy and security. If we do not sufficiently value the airline piloting profession and future pilots are less experienced and less skilled, it logically follows that we will see negative consequences to the flying public ...*

While Colgan Air can be criticised for its lack of interest in pilot welfare, the airline's culture[18] and actions must be considered against a broader tableau. They must be considered *in context*. Thanks to deregulation most airlines are locked in a dog-fight for survival. Robert Crandall, CEO of American Airlines, claimed the industry was 'intensely, vigorously, bitterly, savagely competitive' (Crandall cited in Petzinger 1995). Donald Burr, CEO of pioneering US low-cost carrier People Express, considered airlines 'a low-end commodity … capital intensive, labour-intensive, fuel-intensive' (Burr cited in Sampson 1984). For Burr there was no difference between aviation and the economically unstable smoke-stack industries (steel-making, ship-building, car manufacture, coal-mining). Michael O'Leary, CEO of successful low-cost carrier Ryanair, complained: 'Airlines are very hard to run …' (O'Leary cited in Ashcroft 2004).[19]

As to finding a practical solution to the problem of long-distance commutes to work, a model exists in the form of the United Kingdom government's Key Workers subsidised housing scheme (Direct.gov 2010). The scheme, which acknowledges the high cost of living in or near major conurbations, helps those who provide essential services to buy or rent a home. It helps nurses, firefighters, prison officers, police officers, probation officers, teachers *and* private-sector employees 'who can demonstrate that they are of special value to [the] community and local economy' (Dartford Borough Council 2009) rent or buy good-quality accommodation close to their place of work. This intervention makes the housing market an actant within numerous government and local authority-funded actor-networks (the health care system, the fire and rescue system, the prison system, the education system, etc.). It more closely aligns the property market with key government sectors (and their employees). The Key Worker scheme represents a liberal reading of social responsibility – Keynesianism for the New Millennium.

Given that a significant number of airline managements would probably resist paying towards itinerant pilots' housing costs, it would fall to international bodies (like ICAO and the ILO) and national aviation authorities to establish the necessary conceptual framework, and to national governments and/or local authorities to secure the necessary funds. Of course, the ideal solution would be for airlines to pay all pilots a living wage

---

18 A useful definition of organisational culture is 'the way we do things around here'.
19 In 2010 O'Leary proposed a single-pilot flight-deck to save money. To deal with the risk of pilot incapacitation a member of the cabin crew could be trained to fly the aircraft, said O'Leary.

– one that allows itinerant pilots to rent good-quality accommodation close to base. Unfortunately, three decades of deregulation and – for many airlines – marginal profitability have taken a heavy toll (Massachusetts Institute of Technology 2011). As the Secretary General of the Association of European Airlines (AEA) remarked at the end of 2009:

> *Portions of our industry are close to collapse ... Steps must be taken now to act positively and decisively to create the conditions under which prosperity can be restored .... (Schulte-Strathaus cited in Dunn 2009)*

The regulation of ticket prices might help restore margins and give airlines more room for manoeuvre.[20] The danger, of course, is that airlines might use the extra revenue not to stabilise and re-model the industry but to subsidise inefficiencies.

Like nurses and firefighters, pilots bear an immediate responsibility for the lives of others. Yet, like nurses and firefighters, many pilots are poorly rewarded for this responsibility.[21] Bennett (2009) alleges illogicality in wage scales:

> *Pilots at regional carriers ... are often paid significantly less than pilots at major carriers ... This differential is justified on the basis that the regional sector is a 'stepping stone' to a career with the majors. While superficially plausible this rationalisation breaks down under scrutiny. Given that all commercial pilots occupy a position of trust, surely the fact that some fly turboprops on regional routes while others fly jets on intercontinental routes should make no difference to their remuneration. Remuneration should reflect not the number of persons for whom one is responsible, but the fact of responsibility ... [A] pilot's responsibilities [do] not vary with the type of aircraft s/he flies or service s/he operates. Commercial pilots' responsibilities are invariable. All pilots must deliver a safe and efficient service, regardless of the type of aircraft they fly or routes or schedules they operate. On a moral level the loss of fifty lives (as happened in the Colgan Air accident) is as regrettable as the*

---

20 The retired CEO of American Airlines, Bob Crandall, has argued for the partial re-regulation of the industry: 'Unfettered competition just doesn't work very well in certain industries, as amply demonstrated by our airline experience .... In my view, it is time to acknowledge that airlines ... are more like utilities than ordinary businesses .... [We could] establish minimum fares sufficient to cover full costs and produce a reasonable return. While I would fully support such an approach, the idea is deeply offensive to those who cling to the belief that the markets can solve everything' (Crandall cited in Foust 2008).

21 Is vocationalism exploited by cost-focused employers?

loss of three hundred lives. Let me put it this way: there is no moral or professional disequivalence between the job performed by, say, a Saab 340 pilot operating an east coast shuttle service and that performed by an Airbus A380 pilot. So on what grounds is the former paid less than the latter? Could it be that the regionals are exploiting an apparent (but not substantive) difference between different types of operation for their own selfish ends? And could it be that the majors tolerate this practice because it helps keep their costs down?

The aviation system is composed of numerous actants – governments, regulators, airlines, employees, passengers, financiers, etc. To make the system safe, each actant must behave in a socially responsible manner. Landlords must provide reasonably-priced, sanitary accommodation. Governments and regulators must address conflicts of interest (is it possible for a regulator to promote both economic efficiency *and* safety?).[22] Remuneration must acknowledge the lived reality of the pilot lifestyle. Pilots who can afford to rent accommodation or take a hotel room in proximity to their base must do so. If sleeping facilities are provided, pilots must use them to counteract fatigue. Passengers must understand the safety consequences of consumer pressure for ever-lower ticket prices. Actants' current behaviour produces resident pathogens. As shown here, when system defences are breached, pathogens manifest as either near-misses or fatal accidents.

References

About.com 2009. *Pilot Fatigue and the Crash of Colgan Air*. Available at: http://airtravel.about.com [accessed: 7 June 2009].

Ashcroft, M. 2004. The Ryan King. *British Industry*, September.

Bartol, K.M. and Martin, D.C. 1998. *Management*. New York: McGraw-Hill.

Barty, J. 2009. Politicians are close to tipping the City over the edge. *The Financial Times*, 16 December.

Beck, U. 1992. *Risk Society: Towards a New Modernity*. London: Sage.

Beck, U. 2009. *World at Risk*. Cambridge: Polity Press.

Bennett, S.A. 2006. *After Hubris, Nemesis: Why Flag Carriers Fail*. Leicester: Vaughan College, University of Leicester.

---

22   It has been claimed that the seeds of Britain's BSE outbreak were sown in the dual responsibility of the Ministry of Agriculture, Fisheries and Food (MAFF). MAFF was responsible for both food production and food safety. Feeding animal remains to cattle (to boost productivity) triggered the outbreak. MAFF no longer exists.

Bennett, S.A. 2008. Defensive capacity: The influence of the facilitation-regulation space. *Journal of Risk Research*, 11(5), 597–616.

Bennett, S.A. 2009. Anatomy of an accident. *The Aerospace Professional*, November, 14–15.

Bloomberg News and The Enquirer 2009. Pilot commute limits eyes to avert fatigue. Available at: http://news.cincinnati.com [accessed: 13 January 2010].

Brewer, W.F. and Treyens, J.C. 1981. Role of schemata in memory for places. *Cognitive Psychology*, 13, 207–30.

British Broadcasting Corporation News 2009. *Pilots Protest Over Flying Hours.* Available at: http://news.bbc.co.uk [accessed: 2 January 2010].

British Broadcasting Corporation News 2011. *Q&A: What's the BA Dispute About?* 8 February. Available at: http://www.bbc.co.uk [accessed: 11 February 2011].

Brookes, A. 2002. *Destination Disaster: Aviation Accidents in the Modern Age.* Hersham: Ian Allan.

Buiter, W. 2008. The City of London can no longer afford the expensive luxury of sterling. Available at: http://blogs.ft.com/ [accessed: 28 March 2011].

Business Link 2010. *Corporate Social Responsibility.* Available at: http://online. businesslink.gov.uk [accessed: 26 September 2010].

Cable, V. 2009. *The Storm.* London: Atlantic Books.

Caldwell, J.A. and Caldwell, J.L. 2003. *Fatigue in Aviation.* Aldershot: Ashgate.

Callon, M. and Latour, B. 1981. Unscrewing the big leviathan: How actors macro-structure reality and how sociologists help them to do so, in *Advances in Social Theory and Methodology: Towards an Integration of Micro and Macro-Sociology*, edited by K. Knorr-Cetina and A.V. Cicouvel. London: Routledge, 277–303.

Centre for Asia-Pacific Aviation 2009. *Global Airline Outlook: Industrial Action the Big Threat to Aviation in 2010.* Available at: http://www.centreforaviation.com [accessed: 30 December 2009].

Certo, S.C. and Certo, S.T. 2009. *Modern Management – Concepts and Skills.* London: Pearson.

Colgan Air 2009. *Frequently Asked Questions – Colgan Air Flight 3407.* Manassas, VA: Colgan Air Inc.

Croft, J. 2010. NASA, easyJet to study commuting, fatigue. *Flightglobal*, 13 May. Available at: http://www.flightglobal.com/ [accessed: 21 October 2010].

Crook, M. 2009. *Submission to the National Transportation Safety Board for the Investigation of Colgan Air Flight 3407 Accident, Bombardier Dash 8-Q400, N200WQ Clarence Centre, New York, 12 February 2009.* 4 December. Manassas, VA: Colgan Air Inc.

Cullen, D. 1990. *The Public Inquiry into the Piper Alpha Disaster.* London: Her Majesty's Stationery Office.

Dartford Borough Council 2009. *Key Worker Scheme*. Available at: http://www. dartford.gov.uk/housing/keyworkerscheme [accessed: 30 December 2009].

Davis, K. 1973. The case for and against business assumption of social responsibilities. *Academy of Management Journal*, 16(2), 312–22.

Dey, I. and Walsh, K. 2010. Bankers to face £1m bonus trap. *The Sunday Times, Business Section*, 10 January.

Dienel, H.L. and Lyth, P. 1998. *Flying the Flag: European Commercial Air Transport since 1945*. Basingstoke: Macmillan.

Direct.gov 2010. *Who Qualifies as a Key Worker?* Available at: http://www.direct. gov.uk [accessed: 2 January 2010].

Dunn, G. 2009. Low-cost carriers on the front foot. *Airline Business*, December.

Evans, D. 2009. Rulemaking to be Watched Closely for Allowing More Potential for Pilot Fatigue. *Aviation Safety Journal*, 7 October. Available at: http://asj. nolan-law.com/2009/10 [accessed: 20 October 2009].

Federal Aviation Administration 2009. *Safety Alert for Operators 09014. Date 09/11/09*. Washington, DC: Federal Aviation Administration.

Federal Aviation Administration 2010. *14 CFR Parts 117 and 121 Flightcrew Member Duty and Rest Requirements; Proposed Rule*. Washington, DC: Federal Aviation Administration.

Fortson, D. 2010. JAL 'about to collapse'. *The Sunday Times, Business Section*, 10 January.

Foust, D. 2008. Bob Crandall: How I'd save the industry. *Bloomberg Business Week*, 11 June.

Freeman, S. 2009a. A crowded hub away from home. *The Washington Post*, 4 August.

Freeman, S. 2009b. Senate hearing examines pilots' living conditions. *The Washington Post*, 7 August.

Freeman, S. 2009c. Loudoun investigates local pilot 'crash pads'. *The Washington Post*, 8 August.

Gander, P.H., Gregory, K.B., Connell, L.J., Graeber, R.C., Miller, D.L. and Rosekind, M.R. 1998. Flight crew fatigue IV: Overnight cargo operations. *Aviation, Space and Environmental Medicine*, 69(suppl. 9), 26–36.

Harvey, G. and Turnbull, P. 2009. *The Impact of the Financial Crisis on Labour in the Civil Aviation Industry*. Geneva: International Labour Organisation.

Hiscott, G. 2009. Alistair Darling welcomes bonus pledge by banks. *The Daily Mirror*, 1 October.

Hopkins, A. 2002. *Working Paper 7: Safety Culture, Mindfulness and Safe Behaviour: Converging ideas?* Canberra: Australian National University.

International Civil Aviation Organisation 2003. *100 Years of Civil Aviation*. Montreal: International Civil Aviation Organisation.

International Transport Workers' Federation 2009. Unions Set Out Challenge on Airline Crisis. Available at: http://www.itfglobal.org [accessed: 30 December 2009].

Kirby, A. 1998. Brent Spar's long saga. *BBC News*, 25 November. Available at: http://bbc.co.uk [accessed: 22 December 2009].

Klein, G. 2007. Corruption and recovery of sensemaking during navigation, in *Decision Making in Complex Environments*, edited by M. Cook, J. Noyes and Y. Masakowski. Aldershot: Ashgate.

Koehler, M., Esquivias, P. and Varadarajan, R. 2009. Cutting to fit. *Airline Business*, November.

Krause, S.S. 1996. *Aircraft Safety: Accident Investigations, Analyses and Applications*. New York: McGraw-Hill.

Langer, E.J. 1989. Minding matters: The consequences of mindlessness-mindfulness, in *Advances in Experimental Social Psychology*, edited by L. Berkowitz. New York: Academic Press, 137–73.

Latour, B. 2005. *Reassembling the Social: An Introduction to Actor-Network-Theory*. Oxford: Oxford University Press.

Law, J. 1987. Technology and heterogeneous engineering: The case of the Portuguese expansion, in *The Social Construction of Technical Systems: New Directions in the Sociology and History of Technology*, edited by W.E. Bjiker, T.P. Hughes and T.J. Pinch. Cambridge, MA: MIT Press, 111–34.

Lowy, J. 2009a. *FAA Inspector Warned of Safety Problems at Colgan*. Available at: http://faa-whistleblower.blogspot.com [accessed: 11 August 2009].

Lowy, J. 2009b. *Buffalo Crash Pilots Discussed Sickness, Low Pay*. Available at: http://www.seattlepi.com [accessed: 5 August 2009].

Massachusetts Institute of Technology 2011. *Airline Industry Overview*. Available at: http://mit.edu/airlines/ [accessed: 6 March 2011].

McGreal, C. 2010. *A Pilot's Life: Exhausting Hours for Meagre Wages*. Available at: http://www.guardian.co.uk [accessed: 13 January 2010].

McRae, H. 2010. A lesson in order from the emerging nations. *The Independent*, 6 October.

Miettinen, R. 1999. The riddle of things: Activity theory and actor-network theory as approaches to studying innovations. *Mind, Culture and Activity*, 6(3), 170–95.

National Transportation Safety Board 2009a. *HP Group Chairman Factual Report. Attachment 1: Interview and Information Summaries, Docket No.: SA-531*. Washington, DC: National Transportation Safety Board.

National Transportation Safety Board 2009b. *Public Hearing in the matter of: Colgan Air, Inc. Flight 3407, Bombardier DHC8-400, N200WQ, Docket No.: DCA-*

09-MA-027: Wednesday, May 13. Washington, DC: National Transportation Safety Board.

National Transportation Safety Board 2009c. *Public Hearing in the matter of: Colgan Air, Inc. Flight 3407, Bombardier DHC8-400, N200WQ, Docket No.: DCA-09-MA-027: Thursday, May 14*. Washington, DC: National Transportation Safety Board.

National Transportation Safety Board 2010. *Loss of Control on Approach Colgan Air, Inc. Operating as Continental Connection Flight 3407 Bombardier DHC-8-400, N200WQ Clarence Centre, New York February 12, 2009*. Washington, DC: National Transportation Safety Board.

Oborne, P. 2010. The coalition speaks with many voices. *The Week*, 2 October.

Pank, P. 2011. Business chiefs call for airport strategy. *The Times*, 29 March.

Pope, F. 2010. BP's Gulf oil spill 'pumped 4.4m barrels into sea'. *The Times*, 24 September.

Petri, T. 2010. Statement from hearing on airline pilot fatigue. Available at: http://republicans.transportation.house.gov [accessed: 21 October 2010].

Petzinger, T. 1995. *Hard Landing: How the Epic Contest for Power and Profits Plunged the Airlines into Chaos*. London: Aurum.

Ray, L. 2010. Losing its luster. *Air Transport World*, 1 May. Available at: http://www.atwonline.com [accessed: 27 May 2010].

Reason, J. 1990. *Human Error*. Cambridge: Cambridge University Press.

Risher, W. 2009. Pinnacle pilots picket outside stockholder meeting. *Memphis Commercial Appeal*, 21 May.

Rosenker, M.V. 2009. *Testimony of the Honourable Mark V. Rosenker, Acting Chairman, National Transportation Safety Board, Before the Subcommittee on Aviation Operations, Safety and Security, Committee on Commerce, Science and Transportation, United States Senate. Aviation Safety: FAA's Role in the Oversight of Commercial Air Carriers, June 10, 2009*. Available at: http://www.ntsb.gov/speeches/rosenker [accessed: 21 June 2009].

Ryley, S. 2009. Crash pads for pilots. *New York Post*, 26 July.

Sampson, A. 1984. *Empires of the Sky: The Politics, Contests and Cartels of World Airlines*. London: Hodder and Stoughton.

Schwartz, N.D. 2008. Suddenly, Europe looks pretty smart. *International Herald Tribune*, 20 October.

Simms, A. 2010. BP and the fall of Britain's business empire. *The Big Issue*, 27 September–3 October.

Sochor, E. 1991. *The Politics of International Aviation*. London: Macmillan.

Stalder, F. 1997. *Actor-Network-Theory and Communication Networks: Toward Convergence*. Available at: http://felix.openflows.com/html/Network_Theory.html [accessed: 14 May 2009].

Sullenberger, C.B. 2009. *Statement of Captain Chesley B. Sullenberger III, Captain, US Airways Flight 1549, before the Subcommittee on Aviation, Committee on Transportation and Infrastructure, United States House of Representatives, February 24, 2009*. Washington, DC: United States House of Representatives.

The New York Times 2009. Pilots and fatigue. *The New York Times*, 17 June.

The Spectator 2009. Battle for the city. *The Spectator*, 5 December.

Turner, C. 2010. Colgan Air accident. E-mail communication to author at sab22@le.ac.uk, 6 January 2010, 16:29.

United States Bureau of Labour Statistics 2008. *Occupational Employment Statistics. May 2008 National Occupational Employment and Wage Estimates, United States*. Available at: http://data.bls.gov/cgi-bin [accessed: 6 September 2009].

Wald, M.L. 2009. Pilots set up for fatigue, officials say. *The New York Times*, 14 May.

Weick, K.E. and Sutcliffe, K.M. 2001. *Managing the Unexpected: Assuring High Performance in an Age of Complexity*. San Francisco, CA: Jossey-Bass.

Woollacott, M. 1998. Risky Business, Safety, in *The Politics of Risk Society*, edited by J. Franklin. Cambridge: Polity Press, 47–49.

Zremski, J. and Schulman, S. 2009. How safe are regional airlines? *The Buffalo News*, 18 May.

7

Investigating Resilience, Through 'Before and After' Perspectives on Residual Risk

Hugh Deeming, Rebecca Whittle[1] and Will Medd

Introduction

Flooding is not generally regarded as being the kind of hazard that is symptomatic of a 'Risk Society' (Beck 1992), in which dangers arise as unintended by-products of technological modernisation and an unquestioning faith in the ability of science to solve social and environmental problems. However, this chapter explores policy change and the results of two research projects conducted with flood-exposed and affected communities, to argue that the recent shift towards the Flood Risk Management (FRM) approach, with its associated shift of responsibility towards the individual, is indeed, an example of the Risk Society at work. In short, decades of support for structural solutions, combined with the increasing challenges of climate change, have allowed the expansion of communities into flood-prone areas, thus increasing the risks to individuals when these defences fail. The research results we present here illustrate how the government's policy of 'Making Space for Water' (Department for Food and Rural Affairs 2005) is played out in practice, with consequences for how risk and resilience is experienced by the communities concerned. We conclude by arguing for citizens to be more involved in the decisions that are made around flood risk management and for better support for the process of flood recovery.

---

1   née Sims.

From Flood Defence to Flood Risk Management: Policy Change from 1940 to the Present Day

The relationships between exposure, vulnerability and resilience to hazards have been much debated (e.g. Adger 2000, Adger 2006, Birkmann 2006, Hewitt 1997, Pelling 2003a, Wisner *et al.* 2004). In relation to flooding, these concepts have been used to describe the changing macro- and micro-social and political processes that have guided the human development of floodplains. As far back as 1945, Gilbert White called for a critical examination of the assumptions being made in relation to how 'adjustment measures' were being used to justify floodplain encroachment (White 1945). His concern, even then, was that some floodplains in the US were being used in ways that increased the exposure of communities living in low-lying areas to flood hazards, thereby exacerbating flood risks. Yet floodplain development has continued into the twenty-first century; a phenomenon that remains accountable to the legacy of historical decisions, which initiated and then normalised such practice long before White wrote his thesis (Doe 2006). In the UK, this could be argued to have occurred largely because the benefits of using this land have continued to be perceived to outweigh the costs, of either mitigating the most frequent hazards, or of suffering the consequences of the more infrequent extreme events. For a relatively small, densely populated island, part of this benefit/cost equation undoubtedly relates to the fact that floodplains represent such a large proportion of useful and useable land (Kelman 2003).

Johnson *et al.* (2005b) identify three phases of flood management within England and Wales, which illustrate a gradual progression of policy priorities since the mid-twentieth century. The first phase followed severe fluvial flooding in 1947 and the east-coast storm surge of 1953, which both affected agricultural yields, even if only for a relatively short time (Johnson *et al.* 2005a). This loss of production represented a substantial risk to food security, whose sensitivity to perturbation had already been severely tested during the war years. The first phase of flood control, therefore, ran from the time of war and post-war austerity in the 1940s to the 1980s, with activity during this phase concentrated on land drainage in support of agricultural productivity.

Due to the changing role of global markets, from the 1980s to the 1990s, a reorientation occurred. This second phase refocused attention away from agricultural productivity and towards assuring the nation's wider economic security. This shift was designed to enable economic growth and social-welfare improvements to be driven by the urban and commercial development of the nation's floodplain, with hard-engineered measures being used to prevent inundations. Unfortunately, however, it could be said that this was the period that raised the paradox of flood control into sharp relief. Parker (1995) christened a major part of this paradox the 'escalator effect'. This concept was predicated upon the observation that this type of adjustment measure meant that the investment in flood defences led to increased investment in floodplain development, which in turn led to the need for more investment in flood defence; thus 'escalating' risks. By the 1990s this approach was beginning to attract criticism, with increasingly clarion calls being directed at *'engineering hubris, disaster-denial mentality and a willingness to pursue short-term profit in the face of long-term risk'* (Mount 1998) bearing more responsibility than any increase in hazard frequency for the rising flood losses that were being experienced (e.g. Barredo 2009). It was becoming clear, therefore, that White's words had been prophetic, and that there was a need to reconsider society's relationship with floods.

Another influence that was driving this tension was the growing consensus amongst the clear majority of scientists that something was happening to the global climate. By 2007 this consensus was to develop sufficiently for it to represent an understanding (with more than 90 per cent confidence), that human activity had been the dominant cause of observed climate changes in the latter part of the twentieth century (Intergovernmental Panel on Climate Change 2008). It was this growing recognition of climate change that was to foment the development of the climate-model projections (e.g. Hulme *et al.* 2002, Jenkins *et al.* 2009), which were to inform the third phase of flood management: the FRM approach (Tunstall *et al.* 2004). In fact plans developed through the use of these climate and catchment models serve to epitomise the FRM approach as it is now practised in the UK, across Europe and in the US (European Commission 2006, Defra 2005, United States Army Corps of Engineers 2009).

The significance of this shift to FRM is that it recognises that not all floods can be prevented. Consequently, it reinforces the need to better understand the ways in which different social and physical interventions can contribute to improving flood resilience[2] and resistance, through achieving the combined social, economic and environmental goals inhered within the wider sustainable development discourse (Water Directors of the European Union 2004). Flood policy is now orientated towards the communication of risk information to the population, and on working in partnership with both the hazard-exposed and

---

2   For a discussion of resilience as a concept, which can variously refer to: resistance; bounce-back; adaptation; or transformation, see Whittle *et al.* (2010).

the greater society, towards building resilience across scales, whilst accepting that this resilience *will* be tested. In the UK this approach has been rather euphemistically termed 'Making Space for Water' by the Department for Food and Rural Affairs (Defra 2005).

Despite this change of emphasis away from hard engineering solutions, in England and Wales, expenditure on flood defence is currently higher than at any time in history (Environment Agency 2009) whilst the residual risks associated with the range of flood hazards are also recognised as being larger than ever before (Association of British Insurers 2006, Association of British Insurers 2007, Risk Management Solutions 2003). One consequence of this is that FRM approaches now incorporate a shift of responsibility towards individuals, which proposes that, 'where appropriate', individuals, households and social networks should be encouraged to mitigate their own risks autonomously. People are effectively being encouraged to 'know' the risks they face and to take personal responsibility for adapting to those risks (Water Directors of the European Union 2004). This is a laudable goal. However, it should be remembered that the prevailing structure, transparency and participatory openness of FRM institutions, and the accent of the policies by which they are bound, cannot be thought of without also considering the legacy of earlier approaches, which had a role in determining the nature of both a population's vulnerability and its capacity for resilience (Wisner 2001). The next section seeks to delineate why this legacy of policy changes means that the public cannot simply be expected to accept such an individualisation of flood risk responsibility on *trust*.

Floods and the Risk Society

In his original *Risk Society* thesis, Beck (1992) suggested that the pervasive effects of globalisation, social reflexivity and the onset of a post-traditional social order have combined to create a society of individuals who regard the post-industrial world with doubt, reflexivity and anxiety. Beck's work is more often associated with 'technological' rather than 'natural' hazards (Walker *et al.* 2010) and, therefore, floods are not a 'typical' case. However, as we illustrate here, the instrumentalist ways in which floods have been managed across preceding decades has had the unintended consequence of increasing the risks to life and property when flood defences fail. This, together with the increase in extreme weather events which constitute the projected impact of human-induced climate change, means that floods cannot be understood as purely 'natural' hazards. Viewed in this way the most recent iteration of FRM provides us with a clear example of how Beck's thesis can be applied to certain aspects of flood risk.

The proliferation of floodplain development and the reliance on technical risk-assessment techniques in the construction of defence measures (e.g. benefit/cost analyses) has, in effect, exposed situated and 'trusting' publics to potential harm. To illustrate this point, think about the residents of a new bungalow development on the coast: surely these people should be able to trust that the same authorities who granted planning permission for their homes to be built will also bear the lion's share of the responsibility for preventing those same properties from flooding? And yet flood defences do fail, often with disastrous consequences, as the summer floods that affected the UK in 2007 made only too clear (Pitt 2008) (Figure 7.1).

Figure 7.1   An example of the 'stripping out' process endured by many residents in Hull after the 2007 flood: Where the affected property has been stripped back to its bare structure in order to facilitate repair

Note: In this image the flooring has already been replaced.

Source: © Beccy Whittle.

From this perspective, the threat from flooding can be seen as analogous to the persistent threats from nuclear, chemical and GM technology and the other global threats around which the risk-society thesis is constructed. One could, for example, regard Parker's 'escalator effect' (1995) as one of Beck's 'residual' risks in microcosm; in that it is a vestige of a particular industrial era, when hubris dictated that all flood hazards could be controlled. Use of this lens facilitates an interpretation that the recent policy shift towards greater individual responsibility for personal flood-risk management could leave the public startled at the reflex-like nature of the policy re-orientation. Such an analogy fits well with Beck's own definition, as an example of entry into Risk Society occurring:

> ... at the moment when hazards which are now decided and consequently produced by society undermine and/or cancel the established safety systems of the provident state's existing risk calculations. (Beck 2000: emphasis in original)

Fleshing out this example, policy-makers conferring the new FRM mandate could be perceived as effectively reneging on the state's historical and socially-deemed responsibility for flood-risk mitigation, at the very time when the global threat of climatic instability is projected to intensify future flood hazards and/or make them more frequent (Alcamo et al. 2007, Evans et al. 2008). Society, from this perspective, could be seen as being left atomised in the face of uncertain, and perhaps indeterminate, levels of residual risk (i.e. that which remains after the insurance provided by 'established safety systems' is overwhelmed).[3]

Having set the scene as regards the recent evolution of FRM approaches, the discussion will now move on to investigate how risk-society reflexivity could be said to be exhibited by populations living on particular floodplains in England. This will be done through the interrogation of data from two recent research projects that purposively engaged populations exposed to or affected by surface-water and sea flood hazards.

The two projects took place between 2007 and 2009 and, as the following descriptions illustrate, these were separate studies, which had different aims and methodologies. However, taken together, the projects provide us with an

---

3   This does not mean insurance in the sense of private policies provided by the insurance industry, but the insurance inhered within state-funded risk mitigation measures (e.g. seawalls built to a 1:200 standard of protection, wherein the residual risk is defined as that which would be realised if the structure was subjected to, for example, a 1:400 hazard event).

interesting picture of how flood risk is individualised, perceived and managed. Indeed, by exploring the results from both projects, we can follow the ways in which understandings of risk evolve across the hazard cycle, from hazard-exposed populations who have not experienced a flood event (the 'before' project), to those with recent experience of recovering from a major flood (the 'after' project). The following section describes the methodologies of the projects.

The Research Projects

THE 'BEFORE'

During 2007, a research project (Deeming 2008) was conducted in three English coastal towns (Cleveleys, Mablethorpe and Morecambe). All of these towns had a history of sea flooding, although none had suffered a significant event since at least 1990 (Morecambe). Notwithstanding, ongoing concerns over climate change effects related to storm-surge flooding are increasingly suggestive that more intense and frequent sea flooding may occur in the future (East Lindsay District Council 2005, Lancaster City Council 2007, Morecambe Bay Shoreline Management Plan Partnership 1999); therefore, the aim of the project was to investigate how risk perceptions influenced the levels of community resilience to this low-probability but high-consequence hazard in these towns. The research methods consisted of a survey questionnaire, delivered using a random-systematic approach ($n = 343$). The survey included a relatively unusual proportion of open questions, which were designed to draw unprompted opinions and attitudes from the respondents. For example, two questions the respondents were required to answer in their own words were:

- Can you suggest three things which *you* could do if you got a warning that your street (including your home) was going to be flooded in the *next few hours*?

- What, if anything, do you think could be done in [town name] to help the town cope with flooding in the future?

This first method of collecting information from the public was followed up by a series of focus groups conducted with volunteers recruited from respondents to the initial survey (participants = 24). Analysis of the data took a grounded-theory approach, wherein themes were identified from within the rich datasets, and potential causal relationships behind these themes were hypothesised.

THE 'AFTER'

Following the summer floods of 2007, a team from Lancaster University travelled to Hull, where over 8,600 homes were affected and one person died (Coulthard *et al.* 2007b), in order to carry out an 18-month long investigation into what the long-term flood recovery process was like for people (Whittle *et al.* 2010). During the study, the researchers worked with 42 flooded residents using in-depth qualitative techniques designed to capture the recovery process in real time. The methods were based on established techniques that were used successfully in a previous study, which investigated the community's recovery from the 2001 Foot and Mouth Disease outbreak in Cumbria (Mort *et al.* 2004). Upon recruitment, the participants gave an initial semi-structured interview which enabled them to tell their story of the floods so far. At this point they were introduced to the weekly diary booklets that they were encouraged to keep throughout the duration of the project. The diaries started with a few simple 'warm up' questions where participants were asked to rate their quality of life, relationships with family and friends, and health using a simple scale ranging from 'very poor' to 'very good'. However, the main section of the diaries was the 'free-text' part, where they were encouraged to write about their lives that week. To complement the diaries and interviews, the participants also met for group discussions at quarterly intervals during the project where they were able to discuss the issues that were facing them as a group.

The final element of the methodology was a project steering group, which comprised local and national organisations with an interest in flood recovery.[4] During the study the diarists engaged directly with the steering group through a series of group discussions and facilitated meetings, resulting in a high level of impact on policy and practice (Whittle *et al.* 2010). For example, the project was used as a case study during the development of the Cabinet Office's Draft Framework on Community Resilience (Cabinet Office 2010).

---

4   The Hull Study had a steering group comprising the following organisations: Association of British Insurers, Humber Primary Care Trust, Cabinet Office, JBA Consulting, Diarist, Middlesex University, Environment Agency, National Flood Forum, Hull City Council, North Bank Forum, Hull Community and Voluntary Services, University of Cumbria, Hull Residents and Tenants Association, Yorkshire and Humber Neighbourhood Resource Centre.

We now move on to discuss what the projects can teach us about the ways in which the flood risk society could be said to be created, understood and managed.

Results and Analysis: Experiencing Residual [Flood] Risk Society

'BEFORE'

The research project conducted in the three coastal towns was designed with a focus on a particular sea-flood hazard: storm surge. However, from the first stages of survey analysis it became clear that the population in these towns had a wider breadth of interest. For these people, flood hazards could be differentiated into two main types: storm surge and surface water. Storm-surge flooding, which hadn't been experienced for at least 17 years was perceived as an acute threat, whilst surface-water (i.e. drainage excess) flooding was something that was prevalent and even chronic at the street scale. As a result of this, the hazards invoked very different risk perceptions and feelings of personal efficacy and responsibility regarding risk mitigation.

As a useful way to visualise how these exposed publics perceived how different groups of stakeholders quantify (whether personally or within their institutionalised structures) the uncertainties related to the local flooding, Figure 7.2 applies the concept of the 'certainty trough' to show the perception differentiation between the two hazards. The certainty trough concept was originally devised by MacKenzie (1990), who contended that certainty in relation to the use of scientific knowledge can be conceptualised as forming a trough shape, as the knowledge is produced and then utilised by agents progressively further away from this inception point. MacKenzie illustrated his concept by categorising three groups, through which the knowledge passes: namely, the knowledge producers, the programme loyalists, and the alienated.

In Figure 7.2, the solid line describes the public's apparent perceptions of uncertainty, as it relates to surface-water flooding. Here the perception is that formal actors (e.g. Sir Michael Pitt) and flood policy-makers (e.g. Defra) represent the 'knowledge producers'. These actors are perceived to be aware of considerable uncertainty in relation to the prediction of surface-water flood hazards, particularly at the local-scale (Bales and Wagner 2009, Pitt 2008). However, when, in this example, the knowledge moves to the local planning authorities (LPAs) (i.e. the 'programme loyalists'), it is apparent that the public

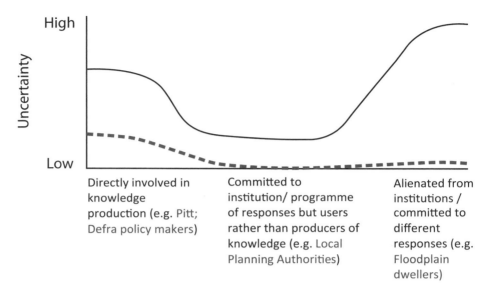

Directly involved in knowledge production (e.g. Pitt; Defra policy makers)

Committed to institution/ programme of responses but users rather than producers of knowledge (e.g. Local Planning Authorities)

Alienated from institutions / committed to different responses (e.g. Floodplain dwellers)

Figure 7.2 **The 'certainty trough' concept applied to the social perceptions of flood-risk management in coastal towns**

perceives that the officers of these institutions justify to themselves that surface-water hazards can be understood with high levels of certainty. The public's perception is that this is what allows the officers to continue to grant planning applications in the towns, without insisting on the implementation of adequate flood mitigation measures. From here the trough is formed, as the knowledge passes to the *'alienated'*. At this final stage the public, who have witnessed chronic flooding at its most localised scale, yet whose 'lay' opinion they feel is rarely sought or considered, perceive the highest uncertainty of all. This perception enhances the feelings of frustration because this is the group that perceives and/or experiences hazards and risks accumulating around them.

Such a situation can result in several responses. The public can:

1. Philosophically accept their situation;

2. Seek to blame others for failing to control surface-water effectively; or

3. Make themselves more resilient.

Whilst the first two options are clearly of limited value in relation to resilience building, the third too should not be considered as being a wholly positive phenomenon. It is true that if a problem is chronic enough and if the person has sufficient perceived self-efficacy – and financial resources – to mitigate the problem (Grothmann and Reusswig 2006), then they can and do install measures such as air-brick covers, sandbags and sump pumps. This is positive, as too is the fact that the data analysis revealed that these actions were often carried out by older people (an oft-cited vulnerable group). However, in this category of uncertainty perception, these actions need to be understood as being undertaken in the sense of (to paraphrase) 'No one else is going to do anything about this so it's down to me!' Whilst the resilience such responses can engender could be regarded positively, it could also be considered as having been attained through rather Machiavellian means. This situation should beg the question: is it ethical to justify such ends (i.e. increased resilience) when, ultimately, the means of achieving it relies, effectively, on the perpetuation of the public's perceptions that risks are being managed ineffectively?

Conversely, the dashed line in Figure 7.2 relates to low-probability storm-surge flooding. Here, the uncertainties inherent in the production of knowledge, as it relates to the prediction of extreme events and to the designed resistance of sea defence measures, is implicit in a lower perceived initial uncertainty being attributed to the 'knowledge producers'.[5] From here the uncertainty perceptions are lowered, as the knowledge passes to local decision makers: the 'programme loyalists'. This is the point at which decisions are written into local development policy. It is, therefore, here that the sustainability of coastal communities is balanced against the risks of a low-probability hazard. In effect, it appeared that local authorities were perceived to *need* to believe that their communities are defended to the highest standard, in order that investment could be attracted and blight avoided. This interpretation was reinforced by a quote from a local councillor who was interviewed as part of the project:

> *At the moment … the Chief Executive and the Leader of the Council both have the attitude that, you know, the sea is there, we're not going to let it come in. We've had hundreds of years of pushing it back, pushing it back, pushing it back, we're holding the line. And they've got to have*

---

5   However, consider Muir-Wood and Bateman 2005) for a discussion of whether low uncertainty, as perceived by the 'knowledge producers' in relation to storm-surge hazards, can actually be equated with low risk. Muir-Wood illustrates that whilst flood probabilities are calculable to an extent and, therefore, uncertainties are relatively low, this does not mean that the residual risks associated with an extreme event are reduced at all.

that story or else the town is so fragile. (Interviewee, Mablethorpe, January 2007)

From here, however, the trajectory of the knowledge diverges from the path taken for surface-water flooding. In this illustration, the 'alienated' publics are also attributed as perceiving low uncertainties. This perception allows people to regard the sea with ambivalence. People know the threat is there (e.g. the North Sea floods of 1953 cannot be denied). Therefore, the sea represents a putative threat. However, the fact that these events are rare suggests that this is a 'low-probability threat'. This message is further reinforced by the words and actions of those who are understood to be responsible for community sustainability (e.g. it is implicit in the decades-long practice of granting planning permission for seaside bungalows). The public are, therefore, able to perceive 'low probability' as meaning that it is not going to happen to 'them' and that if it did then they would be 'unlucky'. Sea flooding becomes an 'Act of God' and sea defences are simultaneously perceived as both impregnable and yet latently vulnerable.

It has been suggested that people with a limited knowledge of certain hazards have a tendency to trust the organisations they deem to be responsible for managing those hazards to mitigate the risks to which they are exposed (Siegrist and Cvetkovich 2000). From a flood-risk perspective, therefore, being able to trust that *someone* is maintaining the standard of sea defences or the drainage infrastructure, or that *someone* will issue warnings in time, allows individuals to perceive that they are exposing themselves to lesser personal risks (Freudenburg 1993). It is this trust that could, in effect, be argued to have produced the perceived division of labour – i.e. the authorities use tax revenue to protect the public, which allows the public to work and pay their taxes in order to sustain the economy – that both cognitively sanctions and perpetuates the risk-taking of those who continue to make the floodplain their home. The public has a vested interest in ignoring even concerted efforts by the responsible authorities in their promotion of the need to build individual resilience to low-probability hazards; regardless of whether such aspirational policy is based on sound science or not. This finding clearly echoes the results from an international research project into social vulnerability:

Most of those surveyed don't feel involved in the decision making processes and tend to delegate responsibility – to agencies in charge of flood prevention and mitigation. Thus, precautionary measures and flood defence are first and foremost regarded as pertaining to public institutions. Such attitudes originate a vicious circle. Public authorities feel the increasing pressure from the residents' demands for assistance and, by positively responding to it, further amplify its magnitude and the citizens' tendency not to invest in prevention. (Steinführer and Kuhlicke 2009: emphasis in original)

Having identified some specific social risk-related phenomena, illustrated by a hazard-*exposed* population, the discussion will now move on to investigate how a hazard-*affected* population was found to rationalise its experience in terms of how risks were realised relative to how they were previously perceived.

'AFTER'

The 'before' case study clearly shows how a risk society is created in relation to flooding. In particular, we can see how the ways in which people understand and manage risks cannot be separated from the macro-scale policy decisions that are made about flooding. The 'after' case study shows these same processes occurring. The Hull flood was a surface water flood and, as a result, there was general agreement from the participants that the city's drainage infrastructure had been found vulnerable. Issues of inadequate maintenance, and failure by local authorities to adequately enforce sustainable drainage management as part of the planning process (and in spite of local protests) were perceived to have exacerbated the consequences. For example:

> *The thing is I mean, the sewerage system is so ancient isn't it? If you are going round the back of Asda, in that area, to my knowledge there's at least five new building [developments], there's David Wilson Homes, there's Persimmon, there's Wimpy's – I don't know how many, all with these new beautiful properties all being built. They were flooded while they were still being built, there's still caravans outside these properties. I mean we complained when they started to build Kingswood, there was a petition up not to build it because you know the properties, there's nowhere for them to go, the drainage system is so old. They are building another, about six companies, are still building in the same vicinity. ('Elizabeth',[6] resident group discussion, 24 April 2008)*

Such comments revealed very clearly that, for many, far from being considered 'Acts of God' these floods were regarded as resulting from 'social' factors; just as were the surface-water hazards on the coast mentioned above. Importantly,

---

6   Diarists' names have been anonymised.

this particular social construction of flood risk can be attributed to one principal factor: namely that the drainage infrastructure in the city had, overall, only ever been required to meet a 'rather vague' industry standard of protection (i.e. one in 30 year:[7] Coulthard *et al.* 2007a). Therefore, an event that achieved a calculated 'greater than 1 in 150 year' intensity (as occurred on 7 June 2007), was inevitably going to lead to extensive flooding. The population of Hull had, effectively, been living with a time bomb of residual flood risk long before those June clouds even formed.

Regardless of this inevitability, however, blame for what had happened led to frustration and anger. Despite the severity of the hazard, physical effects were attributed to inadequate preparation and response by the agencies and organisations. For example:

> *Five years ago the council had decided that they were going to save money and they reduced the drain cleaning from five teams to one team. Now when you look at cities like Rotherham, places like that, they have 20 teams. Well it doesn't take a brain surgeon to see that we have one team. ('James', Group Discussion, Hull, 1 May 2009)*

And:

> *We were like sitting ducks in the middle ... all the buses kept going past and lorries kept going past and they didn't realise that, as they were going past, it was making it like a tidal wave. So it was swishing, and I thought, 'I don't believe this'. Like – with the council – you would have thought between them and the police they would have the sense ... [To close the road]. We've got three main buses that run on that back road. One of them runs every 10 minutes, the other runs every quarter of an hour, so you imagine that every 10 or 15 minutes, what water was getting squished into your house. It was unbelievable. It has been horrendous, really. ('Amanda', Resident Interview, 19 December 2007)*

However, If we follow residents' experiences beyond that of the initial shock and blame and into the longer-term process of recovery, we can see evidence of yet another feature of risk society exhibited by the 'before' research project – namely, the individualisation of risk and its consequences.

---

7   i.e. one chance in 30 in any given year (assuming that probability is constant and that events are independent from year to year: Pielke Jnr. 1999).

Despite the severity of the event itself, the accounts of the diarists showed that most of them were able to deal constructively with their initial situations in ways that enabled them to make a start on the process of recovering their homes. It was what happened next – the struggles with insurers, loss adjusters and builders – that caused real problems for their emotional and mental wellbeing. Indeed, the research shows that the individualised way in which flood recovery is managed, with residents having to deal with as many as 15 different companies and organisations during the repairs to their home, resulted in very mixed experiences for the diarists.

One example of this process was the uncertainty surrounding what would be repaired and how and when these repairs would take place. Even for home owners with full insurance and obvious evidence of water damage there were huge variations in the extent, timing and standards of the work, depending on which insurer the person was with and, in turn, which restoration and building companies were called in. In this way, even neighbours living next door to each other would be treated in completely different ways. These uncertainties were even greater for tenants or those whose homes were affected by 'secondary' flooding – a phenomenon where the water entered below the floorboards, causing structural damage to the home, which was sometimes only detected months later (Whittle *et al.* 2010). In such circumstances, residents had to watch and wait whilst 'experts' from the building, surveying and insurance industries debated the cause and/or significance of damage to their homes and whether their repairs would be covered by their insurance (e.g. was the damage caused by the flood or by the householder's failure to maintain a damp-proof course?). The fact that different 'experts' advised the implementation of different restoration techniques (sometimes in adjoining properties) only added to householders' perceptions of confusion and frustration, particularly if they found themselves unable to influence an insurance company's acceptance of one expert's opinion over another's.

Figure 7.3 provides an example of how it felt for one resident to be an individual trapped in the middle of dealing with all these different companies and agencies. This timeline is taken from the self-assessed scores that Caroline gave herself in the front of her diary during a period from 10 December to 5 May 2008 (see methodology section for more details of this self-assessed scoring process).

As Figure 7.3 illustrates, as well as producing the need to deal with recovery agents, the management of the recovery process was occurring simultaneously

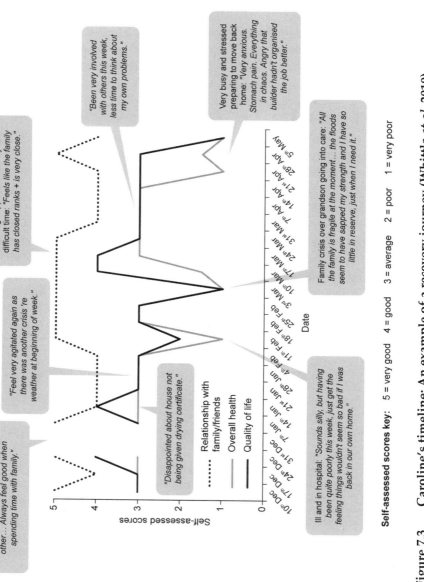

Figure 7.3   Caroline's timeline: An example of a recovery journey (Whittle *et al.* 2010)

with the need to continue everyday life; washing and shopping needed to be done; jobs needed to be held down and dependants needed to be cared for. The 'negotiations'[8] with flood recovery agents merely added to these pressures. This meant that at times some individuals were laden with more responsibility than they felt able to bear. Importantly, however, the research identified that this tendency towards mental and physical overburden did not mean that these individuals were psychologically weak, inherently 'vulnerable' or in some way incapable of dealing with flood hazards. No. These cases of strain were found to relate much more to the fact that people were being *retraumatised* by the way they perceived themselves to be being dealt with by the formal recovery agents. In effect, people who had 'survived' [their expression] the flood, often found themselves strained by the need to generate the additional tenacity that the experience of flood recovery required of them. The following short diary extract illustrates this issue by showing how one diarist described a single exchange with her insurance company, in which she had complained about the service provided by an insurance company-endorsed loss adjuster:

> *Day off today after working on previous Saturday. I call insurance department and speak to them regarding my concerns. I get really upset and have trouble explaining without crying as he says he will call loss adjustor for his side of the story! This comment really upsets me as why should I lie? I insist for his address to post my six page letter and all the copies of emails when [name] has said he will pay rent and storage and never has. I feel absolutely exhausted after this call and feel quite shaken. ('Laura', Diary entry, Monday 23 June 2008)*

Understanding the distress caused by this individualised and longer-term recovery process is important for several reasons: first, for many people, this is how the reality of 'making space for water' is actually experienced in practice, at least at present. It is one thing to ask people to take more responsibility for managing their flood risk but our data shows that, for those who do not know of – or cannot bear the burden of – such responsibility and make the necessary changes to their homes and lifestyles, the consequences can be severe. Second, there are other important longer-term implications of this shift to individual

---

8   'Negotiation' was a phrase that was introduced with some irony by a diarist. He recounted a
    conversation with a loss adjuster during which he realised that householders were not being
    compensated in any uniform manner, but instead whatever 'deal' they received was in fact
    open to negotiation. This is an issue that raises questions about whether those who are less able
    to negotiate for whatever reason would be disproportionately likely to obtain less in terms of a
    final compensation payment than perhaps they were due.

responsibility, which relate to risk and resilience, and we explore these in the following section.

Discussion: Understanding Risk and Resilience

The preceding sections have drawn on data from two research projects to illustrate some of the complexity inherent within the public's multiple understandings of flood risk and responsibility. Whilst Beck's *Risk Society* thesis (1992) might not be immediately considered as applicable, the argument has been laid out in a way that suggests that local flood risks could be understood from a Risk Society perspective, particularly in their terms as drivers of social reflexivity. Using this perspective is not altogether straightforward. It is, for example, true that not everyone is as equally vulnerable to flood hazards as they are to the more traditional of Beck's technological 'icons of destruction' (e.g. nuclear meltdown). In relation to this 'lesser' hazard of flooding some people will inevitably have access to resources (physical, social or financial), sufficient to mitigate their exposure and/or vulnerabilities. However, the experiences of flood-exposed and flood-affected populations reveal some interesting aspects of social reflexivity, which could be suggested to underpin wider public opinions about flood risks and, specifically, who they perceive to be responsible for managing them.

The gradual progression of flood policy in England and Wales over the past century has occurred concurrently to the significant development of many riverside and coastal communities. As a direct result of these policies, the structural defences in all the research sites investigated in this paper are currently regarded as having at least a 1 in 200 standard of sea-flood protection (East Lindsay District Council 2005, Morecambe Bay Shoreline Management Plan Partnership 1999), and in Hull an equivalent main-river hazard protection too (i.e. from the River Humber: Hull City Council 2007). However, due to the legacy of past planning standards and institutional arrangements which could in retrospect be considered inadequate (Environment Food and Rural Affairs Committee 2008), surface-water flood protection standards tend to be lower by almost an order of magnitude.

Yet regardless of what protection standards are designed in,[9] from the perspective of the flood-hazard exposed, physical structures and infrastructure are never 'neutral'. Whether it is a 'nourished' beach, a concrete seawall or a

---

9   Always remembering that a residual risk will remain.

tarmaced-over storm drain, these objectively observable landscape features can act to provide reassurance or concern to these people. Whilst apparently badly-maintained measures (e.g. blocked drains) provide a focus for feelings of blame, reassuring features (e.g. massive seawalls) allow the perception that someone [else] is largely responsible for keeping flood hazards at bay. Surprisingly, the presence of these respective blame and trust-in-authority factors effectively legitimises the household-level investment of emotional and financial capital, in the sense that, to paraphrase, 'someone else is responsible for protecting us'. Thus, houses in at-risk locations continue to be made into homes, and householders' continue to make aspirations for a future *in that place*, without the need to engage with the potentially uncomfortable realisation that they too have responsibility for reducing their own vulnerability.

From this perspective it becomes easier to posit that the shift towards the new FRM paradigm (and its inherent drive towards personal responsibility) could be perceived by this exposed public to be an example of what Beck (1995) would term, the *'organised irresponsibility'* of those who configure FRM's formal institutions. Using this lens, this term could be applied to the process whereby, in acknowledging that floods are too indeterminate to 'know' or to entirely prevent, the FRM authorities have positioned themselves as simultaneously responsible and yet unaccountable. This is a position that allows actors (such as the Environment Agency and local authorities), to be perceived as, on the one hand, taking responsibility for the creation of elaborate 'community' development plans, whilst on the other ceding to the populations an increased personal responsibility for coping with the extreme hazards, which would be capable of sundering any householder's home and/or future aspirations over the course of just a few stormy hours.

Turning the perspective from the exposed towards the flood-affected provides different insights, but ones that are just as relevant to those who seek to reduce future flood impacts by attempting to engage the public with their risks. In the first instance, a kind of resilience was engendered in many residents who acquired new skills as a result of having to 'fight' their corner with the various companies and agencies that they were dealing with. However, as described above, there must be major ethical questions around any form of resilience that has its roots in the unnecessary suffering of residents. Certainly, if the experiences outlined above are anything to go by, then it would be hard to argue that the benefits of such 'resilience' outweigh the cost to the family, individual and community as a whole.

Equally, we can think of the diarists' experiences in the light of Beck's 'safety systems', which allows us to include the consideration of market-based systems, like private insurance. At present, a 'Gentlemen's Agreement' between the government and the insurance industry means that flood insurance will be provided to households exposed to an annual flood probability of no worse than 1 in 75 years until 2013, dependent on the government continuing to invest in flood risk reduction and management measures (Association of British Insurers 2008, Huber 2004). In line with this, prior to the flooding, all the Hull project's diarists would have had access to commercial flood insurance. The fact that most of these people did indeed have such cover was not, however, sufficient for many of them to avoid suffering significant and repeated worry over whether their claims would be met, or whether the future cost of maintaining cover – now that they had submitted a flood-related claim – would remain affordable:

> We went on the web looking for insurances and ... other insurance companies don't particularly want to take you on and the premiums were that high it was unbelievable. So we stuck with the same insurance company and they took us back on and the premium only went up £50 and that wasn't a problem. But the excess has gone up: £5,000 we have to pay on contents and £5,000 on buildings. So if the same thing happened again we've £10,000 to find before we start. And where do we pluck that from? Where do we get that from? We haven't got £10,000. Or do we save anything at all or do we literally just let the whole lot go and say it's all gone and claim what we can and just have everything lesser? ('Leanne', resident group discussion, 17 July 2008)

Leanne's experience provides an opportunity to use Beck's (1992) 'arbiter of risk' characterisation of the insurance industry as an appropriate lens through which to reflect on this particular 'safety system' on offer in the UK. This is because, given its current structure, problems such as rising premiums or large excesses become the inevitable result of the UK's highly individualised, market-driven insurance system, where risks – and premiums – are calculated on a case-by-case basis, with those perceived to be at higher risk (such as those whose houses have been flooded in the past) paying more. In such an arena, insurance availability increasingly becomes a primary concern for anyone living on the floodplain, as well as for those considering buying, selling or developing property in such a place. From this perspective, the development and sustainability of floodplain life could indeed be said to be increasingly contingent upon insurance industry 'arbitration'. Such a system is in complete contrast to France, where solidarity and mutuality are the guiding principles. In France, compulsory cover for disaster risk has been shared since 1982 amongst all policy holders with an identical additional percentage premium paid on top of the assessed premium for fire insurance (French Disaster Reduction Platform 2007).

If the insurance system stays in its present form in the UK then, taking into account the projected increase in extreme weather events resulting from climate change, it stands to reason that the number of people in Leanne's situation will increase every year. Such a scenario is deeply undesirable given that the availability of affordable insurance is often argued to be one of the most important pillars in building resilience for the future (Clark 1998, Pelling 2003b, Whyley et al. 1998) and that the poorest residents will likely be impacted first (Burby et al. 2003).

Insurance also has another important impact on risk and resilience, in the form of its relationship to resilient repair. The Hull study showed that the potential mitigation of future risks fell as a secondary priority to an insurance industry who apparently favoured straightforward restoration over the installation of resilience measures during the repairs (several diarists explained that installing such measures was forbidden by insurance companies as this would have constituted an 'improvement'). For the majority, the insurance 'experience' was one of putting things back as they were, with any incorporation of resilience measures (e.g. concrete floors) resulting more from a particular householder's tenacity or expertise, than because the 'expert' triumvirate of insurer, loss adjuster, and/or contracted builder saw sufficient value in them to insist on their incorporation into the rebuild (see Association of British Insurers 2009).

However, this willingness to forego resilience measures as too expensive was not all the industry's doing. Whilst the principle behind resilience measures was widely accepted by the diarists, their overriding perceptions were that the efficacy of some of these measures was largely uncertain and that any personal expenditure on them might not be recognised in any lowering of the cost of insurance premiums. Accordingly, it became relatively easy for some householders to allow pragmatism to shape their view of a future, wherein, even though the hazard has lost its solely-putative status forever, a refurbished house, bearing no physical reminders of risk levels (e.g. raised electrical sockets), can slowly be returned to a status of home (Harries and Borrows 2007). Through this process, the status quo could be said to be rebuilt into the very fabric of the newly-dried and refurbished buildings, even if it did not quite resettle into the minds and into the 'new normality'[10] of those living in them.

Conclusion: Moving Forward in the [Flood] Risk Society

In this chapter we have argued that the transition to FRM approaches can be viewed as an example of the kind of Risk Society described by Beck (1992). Decades of structural, instrumental approaches to flood defence have resulted in an expansion of homes onto flood plains and exposed coastal areas. We have argued that the creation of such homes, and the engineered structures that protect them, amounts to an unwritten contract which has fostered a belief that science and technology can and should protect us from the dangers that lie on the other side of the flood walls. The large number of severe floods experienced in recent years, however, illustrates that such faith is increasingly misplaced. This has led to a political recognition that escalating expenditure on flood defences is neither financially nor socially sustainable, thus prompting a move to a new approach based on the need to live with flooding through adaptation and resilience. Here, we have argued that a crucial element of this approach involves a shift of responsibility onto individuals. The research results we have described show how this creates a range of ethical and practical challenges for the ways in which risk and resilience are played out, both now and in the future.

So how do we move forward from here? The situation we have described in this chapter is complex and there is little to be gained from attempting to apportion blame to specific individuals or organisations. However, it is possible to suggest a number of potential avenues for change. First, within both studies a particular focus was afforded to the importance of public engagement and to challenging the effectiveness of the *modus operandi* of the recovery organisations and institutions. Regarding the flood-exposed, principal importance was given to the need for iterative engagement processes to be developed, in order that wider introspection can be encouraged on the part of the at-risk publics. Such approaches are also supported by Ronan and Johnston (2005), who promote the idea of 'change-talk'. In a hazards context, this is simply the type of conversation that provokes participants to reconsider their perceptions of what 'safe' really means to them. It effectively provides information that produces a discrepancy between someone's existing understanding of (e.g.) a hazard –

---

10 The phrase 'new normality' was used by a respondent severely affected by the 2001 Foot and Mouth Disease disaster in Cumbria (Mort *et al*. 2005).

as perhaps benign – and the new information that suggests a more malignant presence about which action can and should be taken. However, it is not just the public who need to learn. Effective public engagement allows policy makers and practitioners to benefit from the local knowledge of residents, and an increasing number of research projects point to the importance of creating spaces where residents can get involved in making decisions about their local environment; including those made around flood-risk management (Lane *et al.* in press, Steinführer and Kuhlicke 2009, Whittle *et al.* 2010).

Secondly, if we are serious about making space for water, the Hull study shows that we need to accept that floods will happen and, as a result, we must pay more attention to how people can be supported more effectively during the recovery process. We do not have space here for a full discussion of how to go about this. However, elsewhere we have argued that it is important to address the 'recovery gap', which sees residents trapped in the middle of a very individualising process and having to negotiate themselves through the maze of agencies and companies involved in flood recovery (Whittle *et al.* 2010). We have also proposed that recovery agents, such as the insurance industry, should be considered as more than just market-based businesses. Whether a hazard strikes at the household level or across an entire geographical community, the survivors will need efficient and fair treatment in order that they do not suffer adverse consequences for longer than is necessary. The development of an ethic of care within this particular recovery 'community', through which professional standards could be adhered to throughout the course of any recovery period, should be regarded as vital.

Finally, it is important to consider how our 'safety systems' could be reshaped in order to promote a fairer, more effective sharing of risk and in order to help build resilience for the future. As we have indicated here, this could include revisiting the terms of insurance in order to ensure the continuation of more affordable, equitable premiums, as well as working with insurers and builders to encourage more resilient repair.

References

Association of British Insurers. 2006. *Coastal Flood Risk – Thinking for Tomorrow, Acting Today* (summary report). Available at: http://www.abi.org.uk/ DISPLAY/default.asp?Menu_ID=773&Menu_All=1,773,0&Child_ID=651 [accessed: 7 November 2006]. London: Association of British Insurers.

Association of British Insurers. 2007. *Summer Floods 2007: Learning the Lessons*. London: Association of British Insurers.

Association of British Insurers. 2008. *Revised Statement of Principles on the Provision of Flood Insurance* (Revised 11 July 2008). Available at: http://www. abi.org.uk/Document_Vault/FINAL_INDUSTRY_COMMITMENTS.pdf [accessed: 13 August 2008]. London: Association of British Insurers.

Association of British Insurers. 2009. *Resilient Reinstatement: The Costs of Flood Resilient Reinstatement of Domestic Properties*. London: Association of British Insurers.

Adger, W.N. 2000. Social and ecological resilience: Are they related? *Progress in Human Geography*, 24(3), 347–64.

Adger, W.N. 2006. Vulnerability. *Global Environmental Change*, 16(3), 268–81.

Alcamo, J., Moreno, J.M., Nováky, B., Bindi, M., Corobov, R., Devoy, R.J.N., Giannakopoulos, C., Martin, E., Olesen, J. E. and Shvidenko, A. 2007. *Europe*. In *Climate Change 2007: Impacts, Adaptation and Vulnerability. Contribution of Working Group II to the Fourth Assessment Report of the Intergovernmental Panel on Climate Change*, M.L. Parry, O.F. Canziani, J.P. Palutikof, P.J. van der Linden and C.E. Hanson (eds). Cambridge: Cambridge University Press, 541–80.

Bales, J.D. and Wagner, C.R. 2009. Sources of uncertainty in flood inundation maps. *Journal of Flood Risk Management*, 2(2), 139–47.

Barredo, J. I. 2009. Normalised flood losses in Europe: 1970–2006. *Natural Hazards and Earth Systems Sciences*, 9, 97–104.

Beck, U. 1992. *Risk Society: Towards a New Modernity*. London: Sage.

Beck, U. 1995. *Ecological Politics in the Age of Risk*. Cambridge: Polity Press.

Beck, U. 2000. Risk society revisited: Theory, politics and research programmes. In *The Risk Society and Beyond*, Adam, B.B.U. and van Loon, J. (eds). London: Sage.

Birkmann, J. (ed.) 2006. *Measuring Vulnerability to Natural Hazards*. New York: United Nations University Press.

Burby, R.J., Steinberg, L.J. and Basolo, V. 2003. The tenure trap: The vulnerability of renters to joint natural and technological disasters. *Urban Affairs Review*, 39(1), 32–58.

Cabinet Office. 2010. *Draft Strategic National Framework on Community Resilience* (Consultation Document). London: Cabinet Office.

Clark, M.J. 1998. Flood insurance as a management strategy for UK coastal resilience. *The Geographical Journal*, 164(3), 333–43.

Coulthard, T., Frostick, L., Hardcastle, H., Jones, K., Rogers, D. and Scott, M. 2007a. *The June 2007 floods in Hull Interim Report by the Independent Review Body for Hull City Council*. Hull, UK.

Coulthard, T., Frostick, L., Hardcastle, H., Jones, K., Rogers, D. and Scott, M. 2007b. *The June 2007 floods in Hull Final Report by the Independent Review Body for Hull City Council*. Hull, UK.

Deeming, H. 2008. *Increasing Resilience to Storm Surge Flooding: Risks, Social Networks and Local Champions*. PhD Thesis, Division of Geography Lancaster University, UK.

Department for Food and Rural Affairs. 2005. *Making Space for Water: Taking Forward a New Government Strategy for Flood and Coastal Erosion Risk*. London: Department for Food and Rural Affairs.

Doe, R.K. 2004. Extreme precipitation and run-off induced flash flooding at Boscastle, Cornwall, UK – 16 August 2004. *Journal of Meteorology*, 29, 319–33.

European Commission. 2006. Directive of the European Parliament and of The Council on the assessment and management of floods (2007/60/EC Final). Brussels: European Commission.

Environment Food and Rural Affairs Committee. 2008. *Flooding: Fifth Report of Session 2007–08, Volume 1*. Available at: http://www.publications.parliament. uk/pa/cm200708/cmselect/cmenvfru/49/49.pdf [accessed: 7 May 2008] London: Environment Food and Rural Affairs Committee.

East Lindsey District Council. 2005. *East Lindsey Strategic Flood Risk Assessment (Volume 1 & 2)*. East Lindsey District Council Louth, Lincolnshire.

Environment Agency. 2009. *Flooding in England: A National Assessment of Flood Risk*. Bristol: Environment Agency

Evans, E.P., Simm, J.D., Thorne, C.R., Arnell, N.W., Ashley, R.M., Hess, T.M., Lane, S.N., Nicholls, R.J., Penning-Rowsell, E.C., Reynard, N.S., Saul, A.J., and Tapsell, S.M., Watkinson, A.R., Wheater, H.S. 2008. *An Update of the Foresight Future Flooding 2004 Qualitative Risk Analysis*. London: Cabinet Office.

French Disaster Reduction Platform. 2007. *Insurance and Prevention of Natural Catastrophes, Note by French Delegation to UN-ISDR Global Platform for Disaster Risk Reduction*. Geneva, June 2007.

Freudenburg, W.R. 1993. Risk and Recreancy: Weber, the division of labour, and the rationality of risk perceptions. *Social Forces*, 71(4), 909–32.

Grothmann, T. and Reusswig, F. 2006. People at risk of flooding: Why some residents take precautionary action while others do not. *Natural Hazards*, (38), 101–20.

Harries, T. and Borrows, P. 2007. *Can People Learn To Live With Flood Risk?* Department for Food and Rural Affairs, 42nd Flood and Coastal Defence Conference. York, UK.

Hewitt, K. 1997. *Regions of Risk: A Geographical Introduction to Disasters*. Harlow: Longman.

Huber, M. 2004. *Reforming the UK Flood Insurance Regime: A Breakdown of the Gentleman's Agreement* (Discussion paper 18). Engineering and Physical Sciences Research Council. UK.

Hull City Council. 2007. *Strategic Flood Risk Assessment Halcrow for Hull City Council.* Available at: http://www.hullcc.gov.uk/portal/page?_pageid=221,578 325&_dad=portal&_schema=PORTAL [accessed: 2 March 2010].

Hulme, M., Jenkins, G.J., Lu, X., Turnpenny, J.R., Mitchell, T.D., Jones, R.G., Lowe, J., Murphy, J.M., Hassell, D., Boorman, P., McDonald, R. and Hill, S. 2002. *Climate Change Scenarios for the United Kingdom: The UKCIP02 Scientific Report.* Tyndall Centre for Climate Change Research, School of Environmental Sciences, University of East Anglia, Norwich, UK.

Intergovernmental Panel on Climate Change (ed.). 2008. *Climate Change 2007: The Physical Science Basis. Summary for Policymakers: Contribution of Working Group I to the Fourth Assessment Report of the Intergovernmental Panel on Climate Change.* [Alley *et al.* (Drafting Authors)], Cambridge University Press. Cambridge, United Kingdom and New York, NY, USA.

Jenkins, G.J., Murphy, J.M., Sexton, D.S., Lowe, J.A., Jones, P. and Kilsby, C.G. 2009. *United Kingdom Climate Projections: Briefing report.* Exeter: Hadley Centre, Met Office.

Johnson, C.L., Tunstall, S.M. and Penning-Rowsell, E.C. 2005a. *Crises as Catalysts for Adaptation: Human Response to Major Floods (Report 511).* Middlesex: Middlesex University, Flood Hazard Research Centre.

Johnson, C.L., Tunstall, S.M. and Penning-Rowsell, E.C. 2005b. Floods as catalysts for policy change: Historical lessons from England and Wales. *International Journal of Water Resources Development,* 21(4), 561–75.

Kelman, I. 2003. *Build on Floodplains (Properly). Version 1, 18 September 2003* [Online]. Ilan Kelman. Available at: http://www.ilankelman.org/miscellany/ BuildOnFloodplains.rtf [accessed: 7 March 2006].

Lancaster City Council. 2007. *Strategic Flood Risk Assessment Lancaster City Council.* Available at: http://lancaster.gov.uk/Documents/Planning/ Background%20Documents/SFRA_September2007_Evidence%20_Base_ nyr.pdf [accessed: 13 September 2007].

Lane, S., Odoni, N., Landström, C., Whatmore, S.J., Ward, N. and Bradley, S. (in press) Doing flood risk science differently: An experiment in radical scientific method. *Transactions of the Institute of British Geographers.* Mackenzie, D. (1990) *Inventing Accuracy: A Historical Sociology of Nuclear Missile Guidance.* Cambridge, MA: MIT Press.

Morecambe Bay Shoreline Management Plan Partnership. 1999. *Shoreline Management Plan Sub-cell 11c: River Wyre to Walney Island (Incl. Lune Estuary to Lancaster).* Morecambe Bay Shoreline Management Plan Partnership.

Mort, M., Convery, I., Bailey, C. and Baxter, J. 2004. *The Health and Social Consequences of the 2001 Foot and Mouth Disease Epidemic in North Cumbria*. Lancaster University, UK. Available at: www.lancs.ac.uk/shm/dhr/research/healthandplace/fmdfinalreport.pdf [accessed: 4 January 2010].

Mort, M., Convery, I., Baxter, J., Bailey, C. 2005. Psychosocial effects of the 2001 foot and mouth disease epidemic in a rural population: Qualitative diary based study. *British Medical Journal* doi:10.1136/bmj.38603.375856.68 (published 7 October 2005).

Mount, J. 1998. Levees more harm than help. *Engineering News Record*, 240(5), 59.

Muir Wood, R. and Bateman, W. 2005. Uncertainties and constraints on breaching and their implications for flood loss estimation. *Philosophical Transactions of the Royal Society A: Mathematical, Physical and Engineering Sciences*, 363(1831), 1423–30.

Parker, D.J. 1995. Floodplain development policy in England and Wales. *Applied Geography*, 15(4), 341–63.

Pelling, M. 2003a. *Natural Disasters and Development in a Globalizing World*. London: Routledge.

Pelling, M. 2003b. *The Vulnerability of Cities: Natural Disasters and Social Resilience*. London: Earthscan.

Pielke Jr., R.A. 1999. Nine fallacies of floods. *Climatic Change*, 42, 413–38.

Pitt, M. 2008. *Learning Lessons from the 2007 Floods: An Independent Review by Sir Michael Pitt: The Final Report*. London: Cabinet Office.

Risk Management Solutions. 2003. *1953 UK Floods: A 50 Year Retrospective*. Risk Management Solutions Inc. Available at: www.rms.com/NewsPress/1953%20Floods.pdf [accessed: 18 May 2005]

Ronan, K.R. and Johnston, D.M. 2005. *Promoting Community Resilience in Disasters: The Role for Schools, Youth and Families*. New York: Springer.

Siegrist, M. and Cvetkovich, G. 2000. Perception of hazards: The role of social trust and knowledge. *Risk Analysis: An International Journal*, 20(5), 713–20.

Steinführer, A. and Kuhlicke, C. 2009. *Communities at Risk: Vulnerability, Resilience and Recommendations for Flood Risk Management* (FLOODsite, T11-07-15). FLOODsite, Centre of Environmental Research, a member of Dresden Flood Research Centre. Available at: http://www.floodsite.net/html/partner_area/project_docs/T11_07_15_Vulnerability_resilience_ExecSum_v2_2_p01.pdf [accessed: 28 June 2010].

Tunstall, S.M., Johnson, C. and Penning-Rowsell, E. 2004. *Flood Hazard Management in England and Wales: From Land Drainage to Flood Risk Management*. World Congress on Natural Disaster Mitigation, 19–21 February 2004 New Delhi, India.

United States Army Corps of Engineers. 2009. *USACE National Flood Risk Management Program Guidance US Army Corps of Engineers*. Available at: http://www.iwr.usace.army.mil/nfrmp/docs/USACE_National_Flood_Risk_Management_Guidance_Letter.pdf [accessed: 2 June 2010].

Walker, G.P., Whittle, R., Medd, W. and Watson, N. 2010. *Risk governance and natural hazards* (CapHaz WP2 report D2.1). Available at: http://caphaz-net.org/outcomes-results/CapHaz-Net_WP2_Risk-Governance.pdf [accessed: 5 July 2010].

Water Directors of the European Union. 2004. *Best practices on flood prevention, protection and mitigation*. Water Directors of the European Union.

White, G.F. 1945. *Human Adjustment to Floods*. PhD, University of Chicago, Illinois.

Whittle, R., Medd, W., Deeming, H., Kashefi, E., Mort, M., Twigger-Ross, C., Walker, G. and Watson, N. (2010) *After the Rain – learning the lessons from flood recovery in Hull, final project report for 'Flood, Vulnerability and Urban Resilience: a real-time study of local recovery following the floods of June 2007 in Hull'* Lancaster University, UK.

Whyley, C., McCormick, J. and Kempson, E. 1998. *Paying for Peace of Mind: Access to Home Contents Insurance for Low-income Household*. London: Policies Studies Institute.

Wisner, B. 2001. 'Vulnerability' in Disaster Theory and Practice: From Soup to Taxonomy, then to Analysis and finally Tool. *International Work Conference*. Disaster Studies of Wageningen University and Research Centre 29/30 June 2001.

Wisner, B., Blaikie, P., Cannon, T. and Davis, I. 2004. *At Risk, Natural Hazards, People's Vulnerability and Disasters*, 2nd edition. London: Routledge.

8

Managing Risks in a Climatically Dynamic Environment: How Global Climate Change Presents Risks, Challenges and Opportunities

Todd Higgins

Introduction

Every day media reports tell us that global climate change is becoming an increasing threat to our and the planet's well-being. As we listen to the likes of former US Vice President Al Gore, we are made to feel that the human race is responsible for global climate change. As Beck points out in his 1998 paper, we substitute our anxieties from what nature can do to us, to what we have done to nature. Thus, we appear to be the villains in the global climate change debate. Yet even some of the scientists who have advocated that the climate is changing are questioning their positions. Professor Phil Jones is one who has withdrawn his advocacy for global climate change citing poor personal organizational skills (Petre 2010). So, is climate change occurring? Is it occurring on a global scale? Are the effects being observed in some areas of the world linked to a global climate shift due to anthropogenic activity, or due to natural rhythms in global climates? What are the risks of acting or not acting? And what are appropriate risk management actions? What opportunities are created by climate change risks? Can businesses, governments and individuals capitalize on the opportunities to forge a more sustainable planet regardless of the change in climate or its root causes?

As this chapter is being written, the Gulf of Mexico oil spill saga is unfolding on a daily basis. British Petroleum is being vilified and many Americans are calling for the current situation to be a wake-up call for moving away from petroleum-based energy sources. Not so much for the potential harm being done to the environment of the Gulf of Mexico and to the people who depend upon the Gulf for their livelihoods, but rather as a way to move us from contributing to further global climate change and greenhouse gas emissions. President Obama, in his speech to the nation on 15 June 2010, insomuch as said that this event was an inflection point for the United States to get serious about moving away from petroleum-based energy sources. The event of the oil spill and the subsequent inability of humans to effectively and quickly respond to stem the flow of oil into the Gulf of Mexico has made this event a disaster that anti-oil and anti-greenhouse gas emissions activists can use as their rallying flag. The activists can demand change by highlighting the risks to the environment and the global climate from the evil practices of oil drilling and petroleum use. People living along the Gulf Coast may disagree with Beck's assessment of the Risk Society. Their concerns are focused on the economics of securing income so that they can feed, clothe and shelter their families – what Beck would call natural concerns. The United States' national focus encompasses the natural concerns of the people who make their livelihood from the technological industries along the Gulf Coast as well as less technological industries such as shrimp harvesting. The focus also encompasses the technological concerns that Beck expresses, where worry shifts as it relates to manufactured uncertainty. Without question, the current crisis in the Gulf of Mexico is manifesting itself into Beck et al.'s (1994) 'life and death politics' for all parties involved; and interestingly, at many different levels and along diverse positions of interest. What is missing from this argument is how to place it in context to the risks associated with global climate change.

When one steps back and looks critically at this oil spill, one should ask the question, what is the impact of this event on global climate change, and how will the repercussions of this event affect global climate change opinion and legislation? My belief is that the repercussions of the oil spill will have a longer and further reaching impact on the global climate change debate than the event has on the long-term health of the Gulf of Mexico and the people of the region. It is likely that, as Berger et al. (2001) state:

> ... what becomes 'official' discourse or storyline is controlled by the exercise of power and influence which creates a framework where certain aspects of a field are identified as the 'official' discourse whereas others remain outside.

In an era of government ineffectiveness, having a major industry to demonize is a gift from the gods. The oil spill event is likely to be interwoven into the debate on global climate change by those who influence the national and international debates, and from the debate will come actions – whether scientifically sound or not.

However, the truth is that the power of nature cannot be underestimated; for it is not what we have done to nature, but rather what nature can do to us. Changes in the climate, regardless of the stimulus of change, may manifest themselves in catastrophic events that will be imposed on the living organisms of the earth. Those who try to blame the human race for the environmental impacts of global climate change and those who paint a catastrophic picture of the results of global climate change fail all living creatures on earth. They fail to address the risk issues, and in a timely manner develop risk management strategies that can be implemented to mitigate the results of global climate change on human and other living populations.

In this chapter, theories of risk, crisis and disaster management in the context of global climate change will be discussed with an emphasis on plants and food crops. I will also discuss views on how politics of the Risk Society has shifted thought on global climate change from the effects that climate change will have on human and other living populations, to how humans have brought climate change upon the earth. Specifically, how the effects of global climate change may be mitigated through adaptation of existing technology and application of focused risk management decision-making, resourcing and implementation. Finally, ideas will be offered to the reader to consider as foundations for developing risk management strategies to seize upon opportunities to mitigate the risk of global climate change and to take advantage of opportunities presented by global climate change. What will not be discussed in this chapter is the validity of the science being used to make us aware that global climate change is taking place, or the rhetoric being used to mobilize humans into reducing their footprint on the earth as a means to stop the continuation of global climate change. Our earth is a dynamic body upon which many changes have occurred throughout its history and these changes were not caused by human activity. The greatest advantage that humans have as we participate in these dynamic events of the earth is our capacity to acquire the knowledge to prepare for and adapt to global climate change. If we are only willing to do so.

Beck (2006) discusses the Risk Society in terms of human worry. He states worries shift from natural concerns to technological concerns in response to manufactured uncertainty. In other words, how the technological and sociological development of humans introduces new risks. Risk managers must endeavour to conceptualize the proportions and consequences of these risks and develop strategies to mitigate them. In essence, our technological advancements increase the risks to society because they increase the pathways that risk can take to generate an effect on society. These pathways become intertwined to form risk networks. Risk networks frequently form complicated risk management systems, where the failure of one pathway may have catastrophic effects on a number of the associated pathways and systems within the network. For example, the recent oil platform fire and subsequent pipeline damage in the Gulf of Mexico has created a number of risk scenarios because of the network formed by the technological advances in offshore oil drilling and the marine and estuary environments that are associated with the risk network. This example highlights Beck's theory of manufactured uncertainty and man's impact on nature. It appears that BP's risk managers failed to fully develop risk control measures to prevent accidents and ensure personnel were competent to respond to a severed pipeline incident (British Petroleum 2010). The same example can also be used to consider the late modern theory of risk management as it relates to developing ideological positions, consulting with experts, setting agendas (prioritizing risk) and developing solutions to mitigate the risks in order of priority.

Ekberg (2007) describes the Risk Society as having six risk parameters: the politics of risk; beyond the presence of risk; different understandings of risk; the proliferation of risk definitions; the reflexive orientation to risk; and risk and trust. Giddens (1999) ties risk to modern capitalism and the risks associated with future profits and losses. But is capitalism sustainable in the long term? This question was addressed by O'Connor (1994) who concluded that the evidence weighs against capitalism being sustainable. Giddens (1999) demonstrates that the risks of capitalism can be a positive. It is the assumption of financial risk that generates wealth. The basic assumption is that a risk (financial etc.) can be managed or a reduction in risk exposure can be obtained through the purchase of insurance. Is this still a valid assumption in the face of global risk events? Beck (1998) believes that:

> ... the world of risk society begins where this calculable model of risk, symbolized through the 'private insurance principle' ends.

Indeed, we insure to assure that we can ensure fulfilment of our obligations; transferring risks to various risk-takers. What happens when insurance cannot provide the capital for the risk we face? Global warming may result in lengthy periods of drought throughout the world resulting in a depletion of global food stockpiles (insurance). Insurance may provide farmers with the funds they require to sustain their farming operation through a drought period, but the stockpile of food reserves is slowly depleted as the length of the drought increases. A prolonged drought might mean that governments can no longer assure constituents that they can provide food to meet their needs. No amount of insurance can create new sources of food. To illustrate this point, a prolonged drought during the 2010 growing season resulted in the devastation of spring wheat and barley in Ukraine, Kazakhstan and Russia. Russia is the world's fourth largest producer, and third largest exporter of wheat. The disaster caused Russia to ban food grain exports until the 2011 crop is harvested (Munro 2010). This disaster caused ripple effects throughout the world among the nations that depend on Russian wheat imports. It may create opportunities for US and Canadian wheat farmers, but at what cost in terms of human suffering? Yet technology can create plants that are better able to tolerate drought conditions and produce crop yields to sustain the world's population.

Ekberg (2007) postulates that modern risk is no longer random, but rather the result of unanticipated and unintended consequences of technological achievement. In some instances, the risk comes not so much from technology but the social acceptance and demand for the technology. Floods serve as an example. Flood control technology was developed to reduce the risk of flooding to low-lying areas. As surrounding areas have grown, streets built and paved and soil compacted, the localized infiltration of water has decreased. The result is more water leaving the local area to arrive at a convergent location faster, and thus increasing the risks of flooding. So the risks come from advancements not directly related to the risk factor in question: the development of surrounding areas (which increases the volume of runoff water to be contained by a given flood control structure) creates the risk. Flooding risks will substantially increase in a climatically changed world. Improved water-management strategies that sequester water closer to its point of origin on the earth's surface are needed.

A subset of risk thought, related to late modern theories of risk, is ecological modernization. Ecological modernization is a relatively new school of thought for managing the ecological and environmental problems we currently face, including global climate change. Ecological modernization shares a fundamental belief with late modern theories of risk that technological solutions will be key components of risk management systems.

How does ecological modernization fit into the climate change discussion, and how can it be used to shape policies for addressing global climate change? Ecological modernization is a shift in social and managerial thought from focusing on the damage we have done to nature, to a process of reverse engineering ecological/environmental risks, and developing mitigation strategies or techniques to reduce or eliminate risks at the source of manufacture. Christoff (1996) states:

> ... the new policy culture and its trends are not always simply or primarily intended to resolve environmental problems. They are also shaped by a contest over political control of the environmental agenda and, separately, over the legitimacy of state regulation (predominately in the English-speaking OECD countries).

Mol et al. (2009) referred to ecological modernization as:

> ... the social scientific interpretation of environmental reform processes at multiple scales in the contemporary world.

Janicke (2007) acknowledges that:

> In general, an environmental problem proves politically less difficult to resolve if a marketable solution exists.

Marketability may also take the form of cost reduction. In the face of increased regulation and regulatory costs, well-managed companies will seek means to reduce regulatory costs. Science will play a central role in shaping both environmental policy-making and compliance/avoidance actions. If the beliefs that environmental stewardship increases costs, and natural resources are free goods to be exploited are dispelled, the holistic value of the resource will be reflected. Then a societal realization (value) can be instilled supporting the wise use of resources and reduction of technological risk manufacture.

Achieving a global paradigm shift to an understanding, acceptance and expectation of holistic approaches to reducing technologically-generated risks will require a sustained, fact-based media campaign. Mol et al. (2009) cite the film An Inconvenient Truth as a:

> ... powerful invitation for individuals to demonstrate their moral concerns and translate these commitments into concrete actions of sustainable citizenship and consumption.

They further discuss the role of consumer (public) awareness in creating momentum for consumer lifestyle changes that promote sustainable habits.

While Mol *et al.* (2009) believe that environmental awareness and activism (consumer discretion) may be as, or more, powerful than regulations and laws, Janicke (2007) has a different view of regulation. While he acknowledges that 'Environmental regulation may create impediments for companies and industries', he notes certain advantages to regulation. According to Janicke, regulations can create or support markets for domestic industries; increase the predictability of markets; level the playing field amongst competitors; and create a situation where customers simply have to accept the changes mandated by the regulations.

Fisher and Freundenburg (2001) point out that even if the global society embraces ecological modernization, it will not be implemented concurrently throughout the world. They cite the example of Japan embracing ecologically modern approaches and simultaneously being the world's second-largest economy. Yet a decade later China, a country which does not appear to embrace the concept of ecological modernization, has usurped Japan as the world's second-largest economy. China has become an economic juggernaut at the expense of the environment and has achieved economic greatness without substantial hue and cry from the global community. What are the messages that China is sending to other developing nations and to the developed nations? It appears that China's political leadership is willing to sacrifice the global environment for the sake of economic power and influence. Perhaps China's economic rise can be traced to Western governments whose economic policies foster open and free trade regardless of global environmental impacts?

Berger *et al.* (2001) suggest that environmental policy is frequently developed by non-governmental players who have the means to wield power and influence. Hence, the policies for mitigating global warming impacts may not represent the best practices for achieving atmospheric greenhouse gas (GHG) reductions, or prepare societies to cope with global climate changes affecting their lives.

The risks of global climate change are many: food production, world health, energy production, national security, and economic development, to list but a few. While the risk categories will be discussed on a global basis, risk categories will be regionalized and even localized based on the specific locale/region's vulnerability to a specific risk factor. What is more, the negative effect of a global climate change risk in one locale may be offset by a positive response to climate change in another locale (Schneider *et al.* 2007). For example, food production may be negatively impacted by rising temperatures in some areas, while the same crops are positively impacted in other areas. Global climate change does not occur as a single incident that results in a long-term alteration of the prevailing climate conditions, but rather as a series of events that take place on a constant basis. Our environment is dynamic. The dynamics of global climate change occur at the nano-climate level and contribute to changes in the micro-climate. Environmental changes at the nano- and micro-climate levels may occur randomly or systematically. They may be cumulative or occur as isolated events that have little or no long-term influence on higher climatic levels. Figure 8.1 shows the hierarchy of climates. Nano-climates can be thought of as individual homes, small streams, wooded areas and commercial buildings. Micro-climates can be thought of as collections of nano-climates, such as metropolitan areas, state parks and forests, highways, rivers and lakes. A macro-climate begins to take on a more regional nature; the Mediterranean region is a macro-climate that encompasses portions of several countries but has similar climatic conditions throughout the region. A kilo-climate has a great deal of climatic variety within its boundaries; a kilo-climate may be looked at in terms of biomes or a country or group of countries that have similarities. At the giga-climate level, a continent or an ocean influence the global climate. The global climate is comprised of a series of giga-climates and is influenced by the dynamics of each giga-climate.

Climate change may or may not redefine the boundaries of the existing climates at each level of the climatic hierarchy. Man's actions are likely to have their greatest direct impact on climates at the nano- and micro-climate levels, and less direct influence on the giga and global climates. The hierarchy of climate fits Beck's theory on Risk Society, in which the risks are generated by man's impact on nature. At the macro-climate level and above, the late modern theory tends to frame risk management actions better than does Beck's theories. Strategic goals and methods to combat the risks associated with global climate change fit neatly into the late modern theory of risk.

Solutions to global climate change emanate from the local level and can result in a global change if adopted and implemented by people throughout the world. The saying 'think globally, act locally' is relevant to our discussion of the risks and opportunities presented by global climate change. From a political perspective, political issues while national or international in scope are really a

Figure 8.1   Global climate hierarchy

series of local issues that collectively form the will of the people. Scientific and political leaders must convince local communities that global climate change is an issue that must be collectively addressed and resolved by the national and international communities alike. Without mobilizing public support neither the risks nor the opportunities presented by global climate change will receive the attention and emphasis necessary. Hence, attacking the problem of global climate change will require paradigm shifts in ideology. It will require the involvement and development of subject matter experts to advise policy makers on the steps to take to reduce or eliminate those human activities contributing most to global climate change, thereby enabling politicians and special interest groups to set agendas designed to influence individuals and local communities on the actions to take to reduce human contributions to climate change. This process follows late modern theory very well.

If one is to believe the entirety of the late modern theory, then one would expect capitalism to find a way to solve global climate change, or at least eliminate further contributions to global climate change from human activities. In fact, we see many established and emerging capitalist endeavours that may reduce greenhouse gas emissions and may have an impact on global climate change. Recycling is an established environmental practice. Recycling of many man-made substances has been undertaken for many decades. Advantages to recycling materials are frequently economic advantages that favour the manufacturer and contribute to creating a profitable company by reusing purchased materials and reducing the need to find a suitable location to dispose of those materials that become waste products. A brewer who sells beer in aluminium cans has an incentive to develop a capability to recycle used aluminium cans to be made into new aluminium cans to avoid the costs associated with mining bauxite ore, smelting the ore, and shipping the aluminium ingots for further production into aluminium cans; each of these steps requires the expenditure of fossil fuels which contribute to greenhouse gas emissions and theoretically to global climate change.

Capitalism is also bringing us improved solar energy technology. The 'villain', British Petroleum, is a leader in the development of solar energy systems. If BP is engaged in solar energy development, then it could be said that their corporate risk managers have assessed the long-term risks of continued reliance upon fossil fuels weighed against the long-term sustainability of the corporation and have concluded that to be a sustainable corporation BP must diversify into other energy sectors. The solar energy industry is in many ways a juvenile industry that is characterized by having a large number of small entrepreneurial companies that are highly leveraged and currently not profitable. The same entrepreneurial spirit characterizes the wind power industry, the hydrogen fuel industry, and to a lesser extent the ethanol fuel industry. Each of these represents a capitalistic approach to addressing both the energy and climatic concerns of a growing group of citizens. Yet, even with the development of these industries, the energy that they are currently capable of producing only offsets a small portion of the energy that supports most human activities. Modern day human activities continue to rely upon the uninterrupted supply of petroleum products and other fossil fuels. Fossil fuels will continue to be critical to the human endeavour for the foreseeable future, and the risks associated with fossil fuel use on a global basis are really associated with the human component of the planet Earth. It is likely that human population will continue to grow for the foreseeable future and that economic development throughout the underdeveloped regions of the world will rely on energy sources that emit greenhouse gases.

Returning to our hierarchy of climates we must focus our attentions and efforts on the nano- and the micro-climates to further assess the risks and opportunities of global climate change. Without having meaningful change at the nano- and micro-climate nodes, influence on the global climate is likely to be unachievable.

Human population increases directly affect the nano- and micro-climates. As populations increase, resource consumption will also increase to a point. The economic laws of supply and demand will temper resource consumption but only after resource production has been expanded and has reached geographical, climatic, economic or genetic limits; or supplies have been exhausted. Recovery of resources that are not economically recoverable today may occur in the future to meet the demands of a greater population, but these recovery efforts tend to be more energy-intensive and follow the law of diminishing returns.

There are many risks associated with global climate change, but ultimately four issues are the keys to managing these risks. These four issues may appear to be an over-simplification of a complex global environment, but they represent the risks and opportunities being presented to humankind by the phenomenon of a dynamic global climate. Heat generation, food production, water availability and clean air are the four keystone issues of global climate change risk management. All other global climate change issues stem from one of these root issues.

Simply stated, each increase in human population requires more food, more fresh water and more energy to sustain a quality of life equal to the quality of life existing in every country on earth today. Following the laws of physics and the second law of thermodynamics, we can expect increased generation of heat at the nano- and micro-climate levels due to human activity as populations grow. If underdeveloped nations seek increased development, increased generation of heat can be expected. Eventually, these increases in heat generation at the lower climatic levels can be expected to affect the macro- and kilo-climates, as they may be doing now. Ultimately the effects of increased heat generation will spread around the globe. Economic development is related to industrial development. Industrialized countries have larger economies, use more energy and emit more GHG. Thus, economic development is a risk factor for global climate change. Heat is produced through any exothermic reaction or series of reactions; it is produced by all of our daily activities. Much of the heat produced is dissipated into the environment and its energy value is lost as it proceeds to cool in the nano- and micro-climates. As our heat generation per cubic foot increases, a larger volume of cool air is required to accept and cool the air containing generated heat. Ultimately this could lead to a rise in ambient air temperature at the nano-, micro- and even macro-climates if heat generation is sufficiently dense. Methods of reducing heat generation and/or capture and use of waste heat are opportunities which can result in a reduction of total heat generated globally. Waste heat capture systems suitable for widespread use at the nano-climate level could result in a significant reduction in energy consumption. Managing waste heat can result in cleaner air and reduced water consumption at the nano-, micro- and macro-climates in areas in which fossil fuels are the primary sources of electrical energy production. Using electricity more efficiently with lower heat generation will also require less energy consumption for the purpose of indoor climate control (air conditioning) which will likewise result in fewer emissions from fossil fuel energy generating sources. Food production will be essential to sustaining terrestrial and aquatic ecosystems under the higher ambient temperatures expected for the global climate. It can be expected that human food production will take precedence. Biodiversity is likely to suffer as the need for food presses more land into agricultural use to meet the needs of a growing global human population. Geographically, humans may suffer due to the effects of global climate change on macro- and kilo-climates that support food production needed to sustain large populations.

Shifts in crop growing areas due to climate changes may not have significant impacts on the production of food crops on a global basis. Many of our staple food crops are adapted to wide geographic and environmental/climatic conditions already. For example, corn is grown throughout the American continent and a warming trend in the North American climate is likely to shift corn production to the north and into the prairie provinces of Canada. The current crops of small grains, canola and flax would likely shift further north to cooler areas where they can be successfully grown. The impacts from this crop migration would primarily be on wildlife through destruction of habitat. Vegetable crops that are produced year-round in semi-arid environments where irrigation water is plentiful may not be able to be grown in these areas due to higher priority uses for water and due to plant responses to higher temperatures or disease. For example, lettuce, spinach, cabbage, etc. bolt (initiate reproductive growth) under high temperature growing conditions, and this may have some impact on global food availability, but the shift in growing regions would certainly have an economic impact on the affected regions.

Wolfe (n.d.) and Baldocchi and Wong (2006) assessed the impacts of climate change on agriculture in the Northeast United States and California. Both reports concluded that there were negative and positive impacts of global climate change on agriculture in their respective regions. Baldochhi and Wong looked at the major crops of California and developed eight climate change scenarios to which the response of each of the major crops of the state was modelled. They determined that a 'moderate' shift in the California climate would produce both negative and positive responses among the major crops; different crops responded differently to the various climate models. Baldocchi and Wong did not address the effects of climate change on animal agriculture in their study, although Wolfe did. Both California and the Northeastern United States are major dairy producing regions. Wolfe believes that increasing temperatures will have a negative effect on cow comfort and thus milk production. Longer periods of warmer weather will reduce the efficiency of animal agriculture throughout the world, but especially in the more developed countries where animal agriculture contributes significantly to both human nutrition and regional economics. Because higher average annual temperatures will also promote increased insect populations, the effects of flies and internal parasites on the efficiency of animal agriculture is a significant risk under warmer climate scenarios. This effect might result in marginal producers leaving the dairy industry and placing economic strain on the industry region-wide. Beef, swine and poultry producers are also at risk of declining efficiency of feed conversion, as well as insect and internal parasite issues. Wolfe, like Baldocchi and Wong, concludes that the 'benign' increases in global temperatures will not result in the unsustainability of agriculture within their respective regions.

Dairy enterprises in many developed countries tend to be crop-based systems, where feed is brought to the cows to maximize the energy available for lactation (milk production). Capper (2009) suggests that dairy operations will need to be pasture-based to achieve meaningful carbon sequestration. Pasture-based dairy operations tend to be seasonal, where cows are milked only during the periods of the year when sufficient pasture is available for grazing. Adopting pasture-based dairy operations on a large scale brings the risk of reducing milk production in developed countries, shifting the geographical areas in which dairy operations are located, and reducing the demand for grain and oil seeds (primarily corn and soybeans). An opportunity presented by moving dairy operations to pasture-based systems is to make more corn available for the production of ethanol or for export to feed humans in countries hardest hit by global climate change.

Forests are another significant natural resource vulnerable to global climate change. Climate change (increasing summer temperatures and an increased duration of elevated temperatures) may change the ecology of both forests and grasslands, promoting less favourable species. The risk of wildfires increases with rising global temperatures, threatening not only the survival of the forests and inhabitants of the forest ecosystems, but also potentially releasing millions of tons of stored carbon into the atmosphere (Anonymous 2004). Models used to predict global carbon sequestration or release suggest the existence of a temperature threshold, below which the Earth's vegetation is an atmospheric carbon sink, and above which the vegetation becomes an atmospheric carbon source due to drought, die-back, and/or fire (Anonymous 2004). For the Earth's vegetation to play a significant role in carbon sequestration, global or regional temperatures must remain below the threshold value. The rate of photosynthesis must be greater than the rate of plant respiration for carbon to be sequestered. Additionally, since wood is harvested as a fuel source for many humans living in lesser-developed countries, the rate of tree harvest must be sustainable. That is to say, the rate of harvest must be less than the rate at which trees are planted and allowed to reach maturity. Forest climate change risks also include the rate of tree growth remaining constant or declining as a result of desertification, while the demand for wood is increasing, as available land suitable for tree production declines. The likely result would be an increase in global deforestation and the loss of a significant number of carbon sinks.

Agriculture and forestry cultural practices are also important for carbon sequestration. Plant respiration returns approximately 50 per cent of carbon assimilated through photosynthesis to the atmosphere (Hillel and Rosenzweig 2009). Much of the terrestrial carbon is sequestered in the soil, primarily in amorphous fixed organic compounds known as humus. Humus is an essential component of fertile soils: providing water retention and supplying nutrients to growing plants, while supporting healthy populations of soil animals whose activities and excrement contribute to improved soil structure. Soil fauna also release CO_2 in the process of breaking down detritus into humus. Humus becomes part of the soil carbon bank and plays an essential role in maintaining a healthy soil environment.

Development and cultivation impact soil carbon banks. Development reduces the area of soil available to store carbon. Buildings, roads and parking lots all prevent the storage of carbon beneath their structures. Cultivation, specifically tillage operations, exposes the humus to oxidation and erosion. The humus content of long-cultivated soils declines steadily until a stable content is reached after approximately 80 years of cultivation; once cultivation ceases, another 80 years are spent building soil humus to the pre-cultivation levels (Hillel and Rosenzweig 2009). Cultivation practices known as minimum tillage or no-till reduce the loss of humus in soils and can build soil carbon levels over time. Unfortunately, these tillage methods rely heavily on chemical pesticide applications, genetically modified organisms (GMOs), crop seeds and large horse power tractors. Hence, minimum tillage and no-till farming are not practiced widely throughout the world.

Substantial research has been conducted on the response of plants to elevated levels of CO2, with sufficient evidence generated to conclude that plants typically increase the rate of carbon dioxide assimilation in a carbon dioxide-enriched environment (Dacey *et al.* 1994). Do all our important plant species respond similarly to CO2-enriched atmospheres? Are there limits to assimilation of CO2 in CO2-enriched atmospheres? Are there undesirable by-products produced by plants growing in CO2-enriched environments? The reality is that plant response to elevated CO2 concentrations is variable, being based on a number of factors, primarily water availability, nutrient availability, physiological factors and duration. These factors can affect plant response to elevated atmospheric CO2 concentrations independently or interactively.

Photosynthesis, the process by which CO2 is assimilated into plants, occurs in all plants, terrestrial and aquatic. Respiration also occurs in all plants and releases a portion of the assimilated carbon dioxide back into the atmosphere. How do elevated CO2 concentrations in the atmosphere affect CO2 assimilation and plant respiration, and what is the net effect of these processes in terms of carbon sequestration?

Phytoplankton are one of the primary carbon sinks in the oceans. Phytoplanktons known as Coccolithopores are responsible for converting dissolved CO2 in ocean waters to calcium carbonate. The scientific community has been concerned that increasing atmospheric CO2 concentrations will result in increased CO2 dissolved in the oceans and that this will lower the pH of seawater through the production of carbonic acid. Several studies on various Coccolithopore species have shown decreased net primary production under lower seawater pH conditions. Yet for at least one Coccolithopore species, *Emiliania huxleyi*, net primary production appears to be increased in CO2-

enriched seawater (Ingesias-Rodriguez *et al.* 2008). The study shows *E. huxleyi* is tolerant of slight pH shifts, indicating that this species has the potential to be exploited to sequester CO2. The risks associated with introducing *E. huxleyi* to non-native environments are both environmental and ecological. *E huxleyi* may not adapt to other environmental conditions, such as ocean temperatures, salinity, etc. Additionally, there is a risk that *E. huxleyi* will adapt too well and change the ecological balance of the indigenous phytoplankton, some species of which may be specific food sources for key species in the food chain.

Plant reaction to elevated atmospheric concentrations of CO2 are not completely understood, but there are some encouraging results. Hillel and Rosenzweig (2009) estimated autotrophic (plant) respiration as releasing 50 per cent of total assimilated CO2 and Gifford (1994) reported that the CO2 release from respiration was between 40–60 per cent of gross ecosystem photosynthetic CO2 fixation. Studies by Gonzalez-Meler *et al.* (1996), Bunce (2001) and Drake *et al.* (1999) suggested gas exchange in plants growing in elevated CO2 concentrations will be reduced, reducing the rate of respiration and consequently, lowering the release of CO2 from plant respiration processes. This suggests a net gain in total CO2 fixed, and carbon sequestration may be possible under elevated CO2 conditions. The long-term impacts of reduced respiration rates and their impact on crop yield, plant growth and survivability, and energy available for transfer to other trophic levels in various plant parts has not been fully studied.

Plants 'breathe' through openings in the lower leaf surface called stomates. Stomates regulate gas exchange and transpiration. Increasing the atmospheric concentrations of CO2 generally reduces stomatal opening (Zhou *et al.* 2009), which increases water use efficiency (WUE). Wheat was shown to improve its WUE when grown in an elevated CO2 environment under drought conditions (Wu *et al.* 2002). As Drake pointed out in the summary of the Stomata 2001 working group, more research is needed at the cellular level to build better models for predicting large-scale plant response to elevated CO2 concentrations.

Genetically engineered crops that are heat tolerant, water conservative or capable of highly efficient CO2 fixation at elevated CO2 concentrations must be developed to maintain global food production at current levels (to prevent global starvation caused by decreased crop yields and increased food prices). Scientists have begun to work on the development of crop species possessing one or more climate adaptation traits. Heat tolerance and water conservation are the fundamental goals of many plant-breeding programmes. Development of crop plants that are capable of removing more carbon from the atmosphere is also highly desirable because, in theory, some of the fixed carbon will be deposited in the harvested portion of the plant. Thus it is likely to increase crop yields while helping to fix more atmospheric carbon.

Plant breeding fits into the late modern theory of risk management because it fits the model of finding capitalistic solutions to today's and tomorrow's problems. However, Almas (1999) points out that the development of new 'super' plants is not without technological risk. What remains to be seen is whether agendas will be set at the national and global levels to promote the development of plant-based solutions to the risks of global climate change. While the development of crops that can remove increased amounts of carbon dioxide from the atmosphere may appear to some to be a logical altruistic goal for major corporations to adopt, the reality is that the costs of developing improved non-commodity crops may limit the investment corporations are willing to make in the development of such plants. In this case the risk being managed is the financial risk, where the capital investment is weighed against the expected reward and the latter is found to be less than acceptable to the corporation's management at this time.

Economic development often results in the conversion of vegetated land into developed land containing less total vegetation. In the current scenario, development of land generally results in less CO_2 being fixed at the nano-climate level than was being fixed in the undeveloped land. The challenge and opportunity for plant breeders is to develop landscape plants that are more efficient in carbon capture than the native plants growing in the undeveloped land. Thus the challenge is to achieve a rate of total carbon capture in the developed nano-climate which equals or exceeds the undeveloped land. Genetic engineering provides the foundation for us to achieve the development of more efficient carbon capturing plants, but questions remain as to the availability of funding to create these plants and what the risks of doing so are.

The solutions to managing or mitigating the risks are reducing CO_2 production (source), or increasing CO_2 capture (sink). The role of the soil and plants in carbon management is becoming increasingly recognized as vital components of GHG management plans.

Elevated CO_2 environments have been shown to alter the nutrient balance of plants and soils. Elevated CO_2 concentrations in the leaf and carbon accumulation in leaf tissue appears to result in a lowering of the leaf mineral content and a decrease in the activity of a key carbon-fixing enzyme (protein) (Porter and Grodzinski 1984).

Nitrogen concentrations in leaf tissue has been shown to decrease in some species of oak (Hungate *et al.* 2006, Stilling *et al.* 2003) and bahiagrass (Newman *et al.* 2006) primarily due to an accumulation of carbon compounds in the leaf. Bahiagrass is a forage grass species. If other forage grass species respond similarly in regards to leaf nitrogen content under elevated CO_2 conditions, the significance for the nutrition of the world's animal populations could be profound. Nitrogen is essential for the formation of amino acids, the building blocks of protein. Lower leaf nitrogen content increases the risks of dietary protein deficiencies occurring in grazing, domestic and wild animals. Protein deficiencies in animals can lead to poor development, reduced conception rates, and reduced milk production. The influence of protein deficiency can progress through the food web; the fates of endangered coniferous species may be affected by the change in the quality of forage grasses. Hungate *et al.* (2006) suggested that limited nitrogen was likely to limit scrub oak production under elevated CO_2; this means a likely reduction in acorn production. Acorns are a high-energy food for many species of wildlife.

The nitrogen effect caused by elevated CO_2 in some species follows Liebig's Law of the Minimum (Liebig 1840). Development (growth) will occur to the extent determined by the least available/abundant factor. When a limiting factor is made more available, further development will be determined by the next limiting factor. In this case, some of the limitations in plant development under elevated CO_2 conditions can be overcome with nitrogen fertilization. To attain optimal growth under elevated CO_2 conditions growth factor (mineral) availability must be optimized. This means additional fertilizer application for cultivated crops; additional fertilization rates brings with it a host of environmental concerns and risks. Recently, Danger *et al.* (2008) called into question the application of Liebig's Law for use in describing community effects. Communities tended to co-limit resource use in their study. Species diversity played a role in community adjustment to the limited resource environment.

The carbon fixation pathways of plants fall into two main categories referred to as C3 plants and C4 plants. C3 plant pathways are less efficient in CO_2 fixation under normal atmospheric CO_2 conditions; C4 plants are more efficient under the same conditions. Research evidence points to C3 plants exhibiting improved CO_2 fixation under elevated CO_2 conditions, and C4 plants exhibiting little or no improvements to CO_2 fixation under the same elevated CO_2 conditions; the exception is with C4 plants grown in elevated CO_2 atmospheres under drought conditions (Leakey *et al.* 2006). Ziska *et al.* (1991) found similar C3/C4 responses to elevated CO_2 by tropical plants. What about corn? Corn (*Zea mays*) is a C4 plant that does not show a significant response to elevated CO_2 under normal soil moisture conditions, but it does show greater CO_2 uptake under drought conditions (Porter and Grodzinski 1984).

Tomato, sugar beet, lucerne, all C3 plants, show an initial increase in photosynthesis under elevated CO_2 conditions, but it then decreases as the plant acclimates to the elevated CO_2 concentration in the air (Yelle *et al.* 1989, Besford *et al.* 1990, Kramer 1981). Cipollini *et al.* (1993) and Johnson *et al.* (2001) found mixed responses to elevated CO_2 in shrubs and scrub oak, respectively.

Rice is a staple grain for many countries. Thailand, currently the world's largest exporter of rice, predicts that climate change will reduce rice yields by 14.2 per cent by 2059 (Buddhaboon *et al.* n.d.). India also predicts a climate change-linked yield decline of nearly 10 per cent in its prime rice-growing region (Mathauda *et al.* 2000). The International Rice Research Institute (IRRI) was reported in a 2005 VOANews.com article as saying that rice production must increase 30 per cent to 40 per cent to meet future world demand. The risks of not developing rice varieties adapted to future growing conditions are high; starvation of much of the world's poor looms on the horizon. If rice production falls in Thailand and India as a result of global climate change, will another region or crop be able to take advantage of the shifting climatic conditions to be able to achieve greater yields which make up for the shortfalls elsewhere?

Coffee production is also vulnerable to the effects of climate change. Increased ambient air temperature results in poorer quality coffee when it is produced under the currently popular (and more efficient) open field cultivation. Traditionally, coffee was an understory plant grown under shaded conditions. The University of Michigan (2008) predicts coffee returning to shade grown cultivation whereas Grainger (2009) suggests coffee production will shift to higher elevations. Baker and Haggar (2007) recommend a series of field trials with a range of shade, irrigation, water conservation, erosion control and other techniques to develop mitigation strategies for future coffee cultivation.

Climate change impacts perennial plants more significantly than it impacts annual or biennial cultivated plants. Corn and soybean producers can shift to adapted varieties. Perennial crops such as coffee, fruit orchards and vineyards, and banana and nut plantations are planted to grow in one place for many years. These crops require several years to reach fruit-bearing stage, and additional years to achieve maturity and full production. They cannot be easily relocated each generation due to the influences of climate change. Climate change may result in fewer perennial plant species used for food due to a reduction of suitable habitat and competition from other food crops. The risk to wildlife habitat is that species competitiveness and richness will fluctuate under elevated CO_2 conditions with the ecology of some regions irreversibly altered. Altered habitats may endanger new species and/or push already endangered species to the brink of extinction (Reekie and Bazzaz 1989).

If plants are so varied in their short- and long-term responses to elevated atmospheric CO_2, are they useful tools in mitigating elevated GHG concentrations? Without a doubt, plants must be a tool used in GHG management. If plants must be used in GHG management, how can we best use these tools?

Green plants need to be incorporated as CO_2 and other GHG sinks at each level of the global climate hierarchy, especially the nano-climate. Capture of GHGs at their point of release may be the most effective means of controlling further increases of GHG in the atmosphere. The strategy for GHG capture must begin at the nano-climate.

Living roofs are another means of improving carbon capture in the nano-climate. Roofing systems that incorporate green plants may not only capture carbon, but reduce the carbon footprint of building roofs; carbon is used in roofing systems. New designs can incorporate living roofing systems, but retrofitting existing roofs to accept living roof systems will be challenging on safety and affordability grounds.

Development of land tends to remove CO_2 sequestering plants and humus, replacing them with something more sterile. Commercial developments tend to clear large areas, and while the building may be 'green' the parking lots are certainly not. How can this nano-climate become a contributor to carbon capture? Many plant species perform well under shaded conditions, allowing vertical development of the space. Vertical development of carbon sinks concentrated in areas of carbon release are likely to capture a portion of the released carbon near the source. Integrating vertical plant community development into commercial and residential development planning would require a small paradigm shift from current thought and an aggressive research programme to identify appropriate plants and cultivation techniques suitable for maximizing carbon capture.

Health Implications of Global Climate Change

The distribution of many pathogenic organisms are influenced by climatic conditions, either directly or indirectly by influencing the range of host populations or transmission vector populations. A shift in global climates resulting in warmer average temperatures will allow the expansion of many disease-causing organisms into areas where their presence was previously controlled by exclusive climatic conditions. These new disease pressures may affect human populations directly and indirectly through their impacts on the plants and animals that humans have come to rely on for food, fibre, and other products essential for modern life.

The World Health Organization has identified several implications of global climate change on public health that will require public health policy changes. These are: intervention to ensure global freshwater resources are safe and sanitary and sufficient to meet global demand; enhancing infectious disease surveillance and response capabilities globally; planning to reduce public health crises in disaster response scenarios; capacity-building for health sector responses; building alliances with agencies whose interests lie outside of the public health sector to promote sustainable development (as a means to prevent or reduce the public health risks of global climate change); and holistically improving environmental conditions that promote health worldwide (Campbell-Lendrum *et al.* 2009).

Jonathan Patz, in a 1998 *Science Daily* news article, stated that the threat of global warming hastening the spread of the viral disease dengue fever into temperate regions was significant. Specifically Patz warned that 'Since the inhabitants of these border regions would lack immunity from past exposures, dengue fever transmission among these new populations could be extensive'. Presently, the geographic range of dengue fever is limited by the range of the mosquito *Aedes aegypti* whose overwintering larvae and eggs are killed by freezing temperatures (*Science Daily* 1998).

In 2008, Patz and his colleagues reported on projected consequences of expected extreme precipitation events on the Great Lakes region of the United

States as a result of global climate change. The location of the Great Lakes makes them susceptible to significant quantities of non-point source pollutants entering the waterways from communities and agricultural activities that surround the lakes and their watersheds. Lake Michigan has a great deal of economic and public health significance to the states of Wisconsin, Illinois, Indiana and Michigan. The Great Lakes are the source of drinking water for more than 40 million people. Its beaches provide leisure time enjoyment to millions of people annually and contribute many millions of dollars to the local economy as part of the tourist industry related to the draw of the lakefront beaches. Contaminants that enter the lakes, often from sewer system overflows in the wake of extreme precipitation events, have resulted in beach closures. The frequency of sewer system compromises which release E. coli and other contaminants into the lake are projected to increase by between 50 per cent and 120 per cent by the end of the century (Patz *et al.* 2008). The public health risks to people dependent upon the Great Lakes are significant and suggest that the mitigation strategy to prevent or reduce the frequency of lake contamination as a consequence of global climate change is dependent upon a multidisciplinary approach that includes infrastructure development, runoff water management and development policies that include provisions for storm water management. The implications are that this scenario will be repeated throughout the world and will likely have even greater public health risks in countries without water quality standards or good public health education programmes.

Clean Air

Greenhouse gas emissions impact air quality. Clean air is essential for combating the anthropogenic influences on global climate change. Air quality affects public health and the health of livestock and plants. Poor-quality air can affect the performance of livestock which decreases the efficiency with which food can be produced for human consumption. Crops can be adversely impacted by poor-quality air as seen by ozone damage to radishes and the effects of acid rain on coniferous forests. In both examples the efficiency of plant growth is compromised by destruction of leaf tissue and thus the rate of photosynthesis declines.

Indoor air pollution contains several GHGs. Some plants are able to filter these pollutants from the indoor atmosphere, fixing them at the source. As building codes in developed countries become more energy-focused, the need to actively manage the indoor atmosphere increases. Indoor plants are effective at this; they also capture CO2 produced by the respiration of the inhabitants of the building. Architectural design changes which promote indoor growth and function of plants are a means to improve indoor and outdoor air quality while capturing GHGs at the source should be encouraged by governmental policies and practices.

The risk of air pollution has decreased markedly in developed countries by the passage of legislation that has driven technology to develop cleaner-burning fuels, engines and motors that are more efficient in transferring energy, systems to reduce the emission of particulate matter and changes in chemical formulations to reduce or eliminate emissions of volatile organic compounds. The transfer of these technologies to developing countries is constrained by their costs and societal attitudes towards paying for pollution prevention. As world population continues to grow, with the preponderance of population growth being in developing countries, new strategies and education programmes to foster the adoption of air pollution control technologies in developing countries is an essential task in combating anthropogenic contributions to global climate change.

Water

Managing water is an essential element in the campaign to adapt to global climate change. The two most publicized risks involving water and a globally warmer climate include sea level rise and widespread drought. The rise in sea level will inundate low-lying areas and will likely cost the relocation of millions of humans. Widespread drought will likely result in famines throughout the underdeveloped populations of the world, and will take its toll on plant and animal populations as well as human populations. Of these two risks drought is the most dangerous and in many respects the least predictable. Sea level rise will occur gradually, giving governments, social scientists, regional and local planners and risk managers time to develop and execute well-thought-out plans in an orderly fashion.

Meteorologists suggest that storm patterns will change in both frequency and severity as a result of global climate change. This means that weather will become somewhat less predictable over a season. Early-season flooding may be followed by mid- or late-season droughts. Both of these events would have a significant impact on crop production practices and crop yields. Thus the food supply risks associated with global warming and water are not just a concern from the perspective of drought, but from a seasonal water management perspective. Having water available at critical times in a plant's lifecycle is essential for ensuring adequate crop yields to support a growing population under increased global temperatures. This means that the risk management strategies related to crop production must include water management strategies, water impoundment structures and irrigation systems.

Conclusions

Global climate change will have significant impacts on agriculture, public health, water availability and global economies. Many global climate change impacts can be mitigated through detailed planning and preparation. The impacts of global climate change will be gradual and somewhat predictable which provides us with a timeline for preparation and action *if* we are prudent. It is clear that it will take a multidisciplinary approach to develop and implement the best strategies for dealing with changes related to global climate change. No one group of scientists, politicians, activists or nation will be able to resolve or mitigate the pending changes independently. Furthermore, it is apparent that there is a great need for global climate change talking points to become more scientific and fact-based, and less fatalistic. Many of the problems we face from a changing global climate are problems that we can resolve through leadership, research and preparation. We will have to resolve global climate change problems by working together as a group of nations adopting common approaches that produce measurable results that can be implemented by all nations regardless of economic status or level of development. Capitalism will remain a key component in developing solutions to climate change impacts, and in so doing capitalism itself may become more sustainable.

Climate change may impact some of the world's staple food crops, but it appears that shifts in growing areas and the development of improved crop varieties will largely mitigate any negative impacts of climate change on commodity grains. Specialty crops, such as coffee, may suffer more from climate change and their global supply may become more limited and costly. Capitalism is likely to help resolve reduced supplies of specialty crops since many multinational corporations are engaged in the production and/or marketing of the specialty crops. Hence, they will have an incentive to find solutions to meet consumer demands.

Water availability and arable land will be the two key elements for determining how well we can adapt our food production to a globally warmer climate. Restrictions on developing quality agricultural land may be required to ensure that future generations can meet the global demand for food for human consumption. Water conservation and water management practices will need to be adapted on an international basis to sustain global food production to meet the demands for food with a growing world population. Developing crop plants that are more drought-tolerant or more efficient in their water usage are two means by which the effects of global warming on crop production can be mitigated. Livestock can also be selected for heat tolerance and water-use efficiency. We will probably see changes in the popularity of specific breeds of domestic livestock. Breeds that are capable of performing well under high temperature conditions and/or breeds that require less water for maintenance are likely to become favoured.

A warming global climate increases disease pressures on humans, livestock and plants. Proactive mitigation strategies will be required to ensure human and livestock populations receive appropriate prophylaxis treatments. Plant breeders and agri-chemists will also need to anticipate plant disease spread and susceptibility, and breed new varieties that are resistant to disease pressures, or chemists will need to develop new families of pesticides to deal with the problems.

Many of the risks associated with global climate change are risks that can be mitigated through the application of science and technology. Ecological modernization appears to be the best-fit school of thought to address the issues of global climate change. Social scientists and physical scientists must work in concert to inform the Risk Society not only of the risks, but of the potential solutions. Social scientists must understand what their physical science colleagues are proposing, and develop strategies to inform and influence the social behaviour of individuals and governments to embrace and adopt the necessary strategies.

References

Adams, R.M., B.H. Hurd, S. Lenhart, and N. Leary. 1998. Effects of global climate change on agriculture: an interpretative review. *Climate Research* 11: 9–30.

Almas, R. 1999. Food trust, ethics, and safety in risk society. *Sociological Research Online*, 4 (3).

Anonymous. 1998. Global warming would foster spread of dengue fever into some temperate regions. *Science Daily*, 10 March 1998. Available from http://www.Sciencedaily.com/releases/1998/03/980310081157.htm [accessed: 4 December 2009].

Anonymous. 2004. Western forests, fire risk, and climate change. Pacific Northwest Research Station, US Forest Service, USDA Issue 6, January 2004.

Anonymous. 2005. World rice production must improve to keep pace with population, climate change. *VOANews.com* 19 October 2005. Available from http://www.Printthis.clickability.com/pt/cpt? Action= cpt &title=world rice+ production+ [accessed: 22 December 2009].

Anonymous. 2008. Green coffee – growing practices buffer climate – change impacts.

Anonymous, 2008. *Managing Climate Change Risks. 2008 Corporate Citizenship Report*. ExxonMobil Corporation. Available from http://www.exxonmobil.com/Corporate/energy_climate_mgmt.aspx. (accessed: 12 September 2009).

Anonymous. 2009. Adaption to climate change by reducing disaster risks: Country practices and lessons. *International Strategy for Disaster Reduction*, United Nations.

Anonymous. 2009. National security implications of global climate change. *Pew Centre on Global Climate Change*. August 2009. Available from www.pewclimate.org.

Baker, P.S, and J. Haggar. 2007. Global warming: The impact on global coffee. *SCAA conference handout*; Long Beach May 2007; final draft.

Baldocchi, D. and S. Wong. 2006. An assessment of the impacts of future CO2 and climate on Californian agriculture. *Climate Change Center* report CEC-500-2005-187-SF March 2006.

Beck, U., A. Giddens and S. Lash. 1994. *Reflexive Modernization*. Cambridge: Polity Press.

Beck, U. 1998. The challenge of world risk society. *Korea Journal*, Winter 1998, 199–206.

Beck, U. 2006. Living in the world risk society. *Economy and Society* 35: 329–45.

Berger, G., A. Flynn, F. Hines, and R. Johns. 2001. Ecological modernization as a basis for environmental policy: Current environmental discourse and policy and the implications on environmental supply chain management. *Innovation* 14: 55–72.

Besford, R.T., L.J. Ludeving, and A.C. Withers. 1990. The greenhouse effect: Acclimation of tomato plants growing in high CO2, photosynthesis and ribulose – 1, 5 –Bisphosphate carboxylase protein. *Journal of Experimental Botany* 41: 925–31.

British Petroleum. 2010. Deepwater Horizon accident investigation report. 8 September 2010.

Bunce, J.A. 2001. Direct and acclamatory responses of stomatal conductance to elevated carbon dioxide in four herbaceous crop species in the field. *Global Change Biology* 7: 323–31.

Buddhaboon, C., K. Kunket, K. Pannangpetch, A. Jintrawet, S. Kongton, and S. Chinvanno. N.d. Effect of climate change on rice production in Southeast Asia: A case study in Thailand. *Thailand Research Fund.*

Buttel, F.H. 2000. Ecological modernization as social theory. *Geoforum* 31: 57–65.

Campbell-Lendrum, D., C. Corvalan, and M. Neira. 2009. Global climate change: Implications for international public health policy. *World Health Organization.* Available from http://www.who.int/bulletin/volumes/85/3/06-039503/en/print.html [accessed: 24 November 2009].

Capper, J.L. 2009. Can carbon sequestration save the planet? *Hoard's Dairyman,* 154 (20), December 2009, 745.

Christoff, P. 1996. Ecological modernization, ecological modernities. *Environmental Politics* 5: 476–500.

Cipollini, M.L., B.G. Drake, and D. Whigham, 1993. Effects of elevated CO2 on growth and carbon/nutrient balance in the deciduous woody shrub Lindera beroin (L.) Blume (Lauraceae). *Oecologia* 96: 339–46.

Coviella, C.E. and J.T. Trumble. 2000. Effect of elevated atmospheric carbon dioxide on the use of foliar application of *Bacillus thuingiensis. Biocontrol* 45: 325–36.

Curtis, P.S., B.G. Drake, P.W. Leadley, W.J. Arp, and D.F. Whigham. 1989. Growth and senescence in plant communities exposed to elevated CO2 concentrations on an estuarine marsh. *Oecologia* 78: 20–26.

Dacey, J.W.H., B.G. Drake, and M.J. Klug. 1994. Stimulation of methane emission by carbon dioxide enrichment of marsh vegetation. *Nature* 370: 47–9.

Danger, M., T. Daufresne, F. Lucas, S. Pissard, and G. Lacroix. 2008. Does Liebig's law of the minimum scale up from species to communities? *Oikos* 117: 1741–51.

Drake, B.G., J. Azcon-Bieto, J. Berry, J. Bunce, P. Dijkstra, J. Farrar, R.M. Gifford, M.A. Gonzalez-Meler, G. Koch, H. Lambers, J. Siedow and S. Wullschleger. 1999. Does elevated atmospheric CO2 concentration inhibit mitochondrial respiration in green plants? *Plant, Cell and Environment* 22: 649–57.

Drake, B.G. 2001. Global change and stomatal research – the 21st century agenda. *New Physiologist* 152: 372–4.

Ekberg, M. 2007. The parameters of the risk society: A review and exploration. *Current Sociology* 55: 343–66.

Feder, W.A. 1978. Plants as bioassay systems for monitoring atmospheric pollutants. *Environmental Health Perspectives* 27: 139–47.

Fisher, D. and W.R. Freundenburg. 2001. Ecological modernization and its critics: Assessing the past and looking toward the future. *Society and Natural Resources* 14: 701–9.

Foster, J.B. 2008. Ecology and the transition from capitalism to socialism. *Monthly Review* November 2008. Available at http://www.monthlyreview. org/081110foster.php [accessed: 18 November 2010].

Fuhrer, J. 2003. Agro ecosystem responses to combinations of elevated CO2, ozone, and global climate change. *Agriculture, Ecosystems and Environment* 97: 1–20.

Giddens, A. 1999. *Risk*. Available from http://www.periwork.com/peri_db/wr_ db/2006_April_12_18_57_11/Risk.html [accessed: 22 December 2009].

Gifford, R.M. 1994. The global carbon cycle: A viewpoint on the missing sink. *Australian Journal of Plant Physiology* 21: 1–15.

Gonzalez-Meler, M.A., M. Ribas-Carbo, J.N. Siedow and B.G. Drake. 1996. Direct inhibition of plant mitochondrial respiration by elevated CO2. *Plant Physiology* 112: 1349–55.

Grainger, S. 2009. Climate change threatens Central America coffee. *Reuters*, 20 August 2009. Available from http://www. reuters.com/assets/print? Aid=USTRE57J4F320090820. [accessed: 22 December 2009]

Hall, D.R. 2002. Risk society and the second demographic transition. *Canadian Studies in Population* 29: 173–93.

Hertel, T.W., A.A. Golub, A.D.Jones, M. O'Hare, R.J. Plevin, and D.M. Kammen. 2010. Effects of US maize ethanol on global land use and greenhouse gas emissions: estimating market – mediated responses. *Bioscience* 60: 223–31.

Hillel, D. and C. Rosenzweig. 2009. Soil carbon and climate change: Carbon exchange in the terrestrial domain and the role of agriculture. *CSA News*, 54(6), 4–11 June 2009.

Hu, Q. and G. Buyanovsky. 2003. Climate effects on corn yield in Missouri. *Journal of Applied Meteorology* 42: 1626–35.

Hulme, M. 2009. The science and politics of climate change. *The Wall Street Journal*, 4 December, 2009. Available from http://online.wsj.com/article/SB30 001424052748704107104574571613215771336.html [accessed: 4 February 2010].

Hungate, B.A., D.W. Johnson, P. Dukstra, G. Hymus, P. Stiling, J.P. Megonical, A.L. Pagel, J.L. Moan. F. Day, J. Li, C.R. Hinkle, and B.G. Drake. 2006. Nitrogen cycling driving seven years of atmospheric CO2 enrichment in scrub oak woodland. *Ecology* 87: 26–40.

Idso, S.B., S.G. Allen, M.G. Anderson and B.A. Kimball. 1989. Atmospheric CO2 enrichment enhances survival of *Azolla* at high temperatures. *Environmental and Experimental Biology* 29: 337–41.

Ingesias-Rodsiguez, M.D., P.R. Halloran, R.E.M. Rickaby, I.R. Hall, E. Colmenero-Hidalgo, J.R. Gittens, D.R.H. Green, T. Tyrrell, S.J. Gibbs, P. Von Dassow, Eirehm, E.V. Armbrust, and K.P. Boessenkool. 2008. Phytoplankton calcification in a high – CO2 world. *Science* 320: 336–40.

Janicke, M. 2007. Ecological modernization: New perspectives. *Journal of Cleaner Production* 16: 557–65.

Johnson, D.W., B.A. Hungate, P. Dijkstra, G. Hymus, and B. Drake. 2001. Effects of elevated carbon dioxide on soils in a Florida scrub oak ecosystem. *Journal of Environmental Quality* 30: 501–7.

Kramer, P.J. 1981. Carbon dioxide concentration, photosynthesis, and dry matters production. *Bioscience* 31: 29–33.

Leakey, A.D.B., M. Uribelarrea, E. A. Ainswaorth, S.L. Naidu, A. Rogers, D.R. Ort, and S.P. Long. 2006. Photosynthesis, productivity, and yield of maize are not affected by open-air elevation of CO2 concentration in the absence of drought. *Plant Physiology* 140: 779–90.

Leakey, A.D.B. 2009. Rising atmospheric carbon dioxide concentration and the future of C4 crops for food and fuel. *Proceedings of the Royal Society of Botany* 7 July 2009 vol. 276 no. 1666 2333–43. Available from http://rspb.royalsocietypublishing.org/content/276/1666/2333.short [accessed: 4 February 2010].

Leiserowitz, A. 2006. Climate change risk perception and policy preferences: The role of effect, imagery, and values. *Climate Change* 77: 45–72.

Makadho, J .1996. Potential effects of climate change on corn production in Zimbabwe. *Climate Research* 6: 147–151.

Martens, W.J.M., L.W. Niessen, J. Rotmans, T.H. Jetter, and A.J. McMichael. 1995. Potential impact of global climate change on malaria risk. *Environmental Health Perspectives* 103: 458–64.

Mathauda, S.S., H.S. Mavi, B.S. Bhangoo, and B.K. Dhaliwal. 2000. Impacts of projected climate change on rice production in Punjab (India). *Tropical Ecology* 41: 95–8.

Mol, A.P.J. and G. Spaargaren. 1993. Environment, modernity and the risk-society: The apocalyptic horizon of environmental reform. *International Sociology* 8: 431–59.

Mol, A.P.J., G. Spaargaren and D.A. Sonnenfeld. 2009. *Ecological modernization: three decades of policy, practice, and theoretical reflection.* Available from http://www.esf.edu/es/sonnenfeld/reader_intro.htm [accessed: 30 December 2009].

Munro, E. 2010. Global trends. *Successful Farming* Special Marketing Issue 2010.

Newman, Y.C., L.E. Sollenberger, K.J.Boote, L.H. Allenge, J.M. Thomas, and R.C. Little. 2006. Nitrogen fertilization affects bahiagrass responses to elevated atmospheric carbon dioxide. *Agronomy Journal* 98: 382–7.

O'Connor, J. 1994. *Is Sustainable Capitalism Possible?* A Guilford Series: Is Capitalism Sustainable? Edited by Martin O'Connor. New York/London: The Guilford Press.

Patz, J.A., S.J. Vavrus, C.K. Uejio, and S.L. McLellan. 2008. Climate change and waterborne disease risk in the Great Lakes Region of the US. *American Journal of Preventative Medicin* 35(5).

Pellow, D.N., A. Schnaiberg and A.S. Weinberg. 1999. Putting the ecological modernization thesis to the test: The promises and performance of urban recycling, in Arthur P.J. Mol and David A. Sonnefeld (eds), *Ecological Modernization Around the World: Perspectives and Critical Debates*. Ilford and Portland, OR: Frank Cass & Company, 109–37.

Petre, J. 2010. Climategate U-turn as scientist at centre of row admits: There has been no global warming since 1995. *Daily Mail online*. Available from http://www.dailymail.co.uk/news/article-1250872/climategate-U-turn-Astonishment-scin [accessed: 17 February 2010].

Porter, M.A. and B. Grodzinski. 1984. Acclimation to high CO_2 in bean. *Plant Physiology* 74: 413–16.

Prios, S.A., H.A. Torbest, G.B. Runion, and H.H. Rogers 2004. Elevated atmospheric CO_2 in agro ecosystems: Residue decomposition in the field. *Environmental Management* 33: 5344–54.

Reekie, E.G. and F.A. Bazzaz. 1989. Competition and patterns of resource use among seedlings of five tropical trees grown at ambient and elevated CO_2. *Oecologia* 79: 212–22.

Rosenzweig, C. and M.L. Parry. 1994. Potential impact of climate change on world food supply. *Nature* 367: 133–8.

Schneider, S.H., S. Semenov, A. Patwardhan, I Buston, C.H.D. Magadza, M. Oppenheimer, A.B. Pittock, A. Rahman, J.B. Smith, A. Suarez, and F. Yamin. 2007. Assessing key vulnerabilities and the risk from climate change. Climate change 2007: Impacts, Adaptation and Vulnerability. *Contribution of Working Group 11 to the fourth assessment report of the, Intergovernmental Panel on Climate Change*, M.L. Parry, O.F. Canziani, J.P. Palutikof, P.J. Vander Linden and C.E. Hanson (eds). Cambridge: Cambridge University Press, 779–810.

Stiling, P., D.C. Moon, M.D. Hunter, J. Colson, A.M. Rossi, G.J. Hymus, and B.G. Drake. 2003. Elevated CO_2 lowers relative and absolute herbivore density across all species of a scrub-oak forest. *Oecologia* 134: 82–7.

University of Michigan, 1 October 2008. Available from http://www.ns.umich.edu/htdocs/releases/planinstory.php? Id=6761 [accessed: 22 December 2009].

Wiedermann, T. and J. Minx. 2007. A definition of 'carbon footprint'. *ISA UK Research Report 7-01*, ISA UK Research and Consulting, Durham, UK.

Wolfe, D.W. 2006. *Climate Change Impacts on Northeast Agriculture: Overview*. Ithaca, NY: Cornell University.

Wu, Q., G-X Wang, Y.F Bai, J-X Lias, and H-X Ren. 2002. Response of growth and water use efficiency of spring what to whole season CO_2 enrichment and drought. *Acta botanica sinica* 44: 1477–83.

Yelle, S., R.C. Beeson, JR., Trudel, and A. Gosselin. 1989. Acclimation of two tomato species to high atmospheric CO_2 *Plant Physiology* 90: 1465–72.

Zhou, H., Y. Chen, W.Li, Y. Chen, and L. Fu. 2009. Photosynthesis of *Populus enphratica* and its response to elevated CO_2 concentration in an arid environment. *Progress in Natural Science* 19: 443–51.

Ziska, L.H., K.P. Hogan, A.P. Smith, and B.G. Drake. 1991. Growth and photosynthetic response of nine tropical carbon dioxide. *Oecologia* 86: 383–9.

9

A Future for Late-Modern Social Formations in Detroit?

Simon Bennett

Introduction

This chapter discusses the concept of ideological latency – specifically the degree to which the Fordist mind-set persists as an interpretative and problem-solving device in so-called post-industrial societies. It explores the experiences of Detroit's urban farmers and small-time entrepreneurs in their efforts to revitalise a severely decayed inner-city economy.[1] It is suggested that, while Detroit's social entrepreneurs embrace such post-Fordist (late-modern) concepts as diversification, flexibility, individuation, improvisation and responsiveness, the city's politicians, bureaucrats, businesspeople and (remaining) industrialists adhere to Fordist concepts like rationalisation, scale, massification, hierarchical management structures, the privileging of expert over other, more grounded forms of knowledge, top-down planning and scientific and technological innovation. It is concluded that (at least in the case of Detroit) the Fordist mind-set or idiom lingers on. Fordism's tangible manifestations – its large-scale production facilities, transport networks and housing projects, and mass-membership social movements – may have been eviscerated by the economic, social and political dynamics of globalisation, but the ideas of Henry Ford endure – the past persists. In Detroit the ideology of Fordism informs City Hall's problem solving: it influences how politicians, bureaucrats and businesspeople frame the problems and potentialities of their city and its people.

---

1   Rebecca Solnit (2007), in her essay Detroit arcadia, describes the landscape of urban Detroit as 'post-American'.

Detroit City Hall's ideological latency is unsurprising. Epochs overlap. They interpenetrate. Progress is messy. To experience progress is to experience ideological, cultural, organisational and technological parallelism:[2]

> [There is a] temptation to exaggerate the new and to represent it one-sidedly, without taking account of the enormous unevennesses and ambiguities that characterise the process of change. (Hall and Jacques 1989)

> Fordism is still alive and well in many places. So, by the way, are pre-Fordist forms. The point is that, despite these many lags and delays which complicate the picture and make definitive assessment difficult, post-Fordism is at the leading edge of change, increasingly setting the tone of society and providing the dominant rhythm of cultural change. (Hall and Jacques 1989)

> [S]ocial development does not fall into neatly demarcated periods. There are very few clean breaks. Attitudes and habits common in one period give way to others only gradually, and many will survive in particular geographical regions and social groups. When sociologists talk about social change in terms of epochs, they are, like caricaturists, identifying and amplifying the most significant features of a society. (Bruce 1999)

Yes, we live in a world where the wealthy are more conscious/less tolerant of man-made risks (Beck 1992, 2009, Giddens 1999) and yes, we live in a world whose culture is homogenising (Jacques 1989). Yes, multi-national corporations (MNCs) wield significant power, and yes MNCs relocate production to low-wage, low-strike economies (Held 1989, Urry 1989). It is fashionable to talk of MNCs' disinvestment in developed nations, of the rise of the BRIC (Brazil, Russia, India and China) countries, of 'economic tigers' (Asian and Celtic), of rust-belts, of service or knowledge-based economies and persistent economic instability (Gorz 2000, New Economy Task Force 2000). But manufacturing still makes a significant contribution to most Western economies, including those of Britain[3] and the United States. It is *wrong* to talk of Detroit as a post-

---

2   Nowhere is this parallelism more apparent than in the religious challenge to scientific rationality, a phenomenon underwritten in the United Kingdom by the government's support for faith schools (some of which fail to challenge creationism). The pre-modern co-exists with the modern.

3   Britain is ranked seventh amongst the world's manufacturing nations. There are 2.5 million manufacturing jobs in Britain: '[T]he UK has significant strengths, notably in pharmaceuticals, aerospace, defence and cars, and in high technology, research and design. Mid-sized companies

industrial city. Manufacturing has declined. But Detroit is still a centre of truck, automobile and war material manufacture, albeit on a reduced scale.[4] The city attracts inward investment (Simon 2010). In Detroit (and Birmingham and Manchester and perhaps elsewhere?) the past is still present.

Globalisation and long-term economic decline[5] have visited numerous ills on Detroit's citizens. Those left in the inner-city[6] experience high levels of crime, poor public services, income poverty, poverty of opportunity, poverty of choice (no national supermarket chain is represented in urban Detroit) and environmental disamenity. The *Risk Society* thesis (Beck 1992, 2009) plays out in citizens' awareness of their plight and direct experience of disamenity and threat (sub-standard housing, a 'crack economy' organised by murderous gangs, limited access to fresh produce, a failing schools system, an overstretched, under-funded police force, an out-of-control arson problem, racial tensions, poor public image, etc.). For urban Detroiters, *Risk Society* is more than a suspicion, more than a perceptual frame. It's a way of life, a *lived reality*.

Motor City Dystopia

> *Practical ideas are the currency of [Detroit] ... Ford was successful because he had 'a better idea', the division of labour, the assembly line, and the vertical factory ... These practical ideas produce their own economy, reinscribe space and transform the city. Each new idea leaves the past in its wake, rendering vast sections of [Detroit] obsolete. (Hoffman 2003)*

---

that have concentrated on global niche markets sense the possibility of some kind of renaissance over the next ten years' (Groom 2010). Britain's nascent space industry directly employs 19,000 people, and is set for further growth (Davis 2010). The 2010 Conservative–Liberal coalition government intends to re-balance Britain's economy in favour of manufacturing. Under the Blair/Brown administrations manufacturing's share of GDP fell from 18 per cent to 13 per cent (Jack 2010).

4   A 2009 University of Michigan study predicted there would be 95,500 blue-collar auto manufacturing jobs across the state by the end of 2011. In 2008 the state had 172,350 blue-collar auto manufacturing jobs (Andersen-Brower and Johnston 2010).

5   The American Institute of Architects (AIA) (2008) notes: 'Between 2001 and 2007, Wayne County [Detroit is located in Wayne County] lost 100,000 jobs in manufacturing, leaving many skilled workers either unemployed or employed in nonmanufacturing (and probably lower-wage) sectors.' In its May 2010 Michigan Forecast the University of Michigan (2010) predicted 'yet another year of job loss, totalling 19,700 workers during 2010 – one-twelfth of the drop posted during 2009'. The University of Michigan (2010) noted: 'Personal income tumbled by 3 percent in 2009, its only yearly decline since at least 1970' and predicted: 'Despite a little stronger growth in personal income in 2011, real disposable income declines by 0.8 percent for that year ...'.

6   The population of Detroit, once around 2,000,000, now stands at roughly 800,000 (American Institute of Architects 2008). Ten thousand leave Detroit each year (Solnit 2007).

> *Detroiters ... like to speak up ... [T]hat feistiness is a sign of their great strength, their great determination to stick it out, to weather the storm, to find solutions. (Babson cited in Cope 2004)*

> *Is perpetual growth the only economic model for cities, or are there also benefits from the de-urbanisation of cities, such as the affordability of spaces and the increase of open land? (Park 2004)*

America's once-thriving motor city, Detroit, has suffered a perfect storm of urban dysfunction, neglect and decay.[7] In 2009 *Forbes* magazine ranked Detroit America's second most empty city, the metropolis having between 60,000 to 80,000 abandoned businesses and homes. Roughly one third of the Detroit city area lay derelict.[8] The median household income of Detroit's residents had dropped 24 per cent in eight years. Detroit's school system was in a state of chaos, with frequent changes of superintendent and high dropout rates. A local authority spokesperson commented: 'We are losing a generation of kids' (Patterson cited in Billups 2009). In 2009 one half of Detroit's children lived in poverty. The literacy rate for the city's adults was less than 50 per cent. Lacking skills, education and disposable income Detroit's residents held little appeal for employers or businesses. The drift to the suburbs and satellite towns continued. Detroit risked becoming an empty husk. It also risked bankruptcy. Fewer businesses meant lower tax revenues, which meant higher taxes for its remaining citizens, who, lacking jobs, were less able to pay. In 2010 Detroit's

---

7   Detroit is not alone in experiencing severe social and economic dislocation. Gary, Indiana, has experienced similar difficulties. Home to US Steel Corporation's biggest mill, the city has seen its population halve: '[T]he salariat, including many of the steelworkers, has moved out, or moved into landscaped and patrolled communities on the edge of town' (Mason 2010). This 'hollowing out' has left Gary's downtown area almost deserted by day. Crime is rampant. Gary created 327 jobs with $266 million of fiscal stimulus money (each job cost $800,000). One interpretation of this lack of success is that in the United States (and the United Kingdom?) only the private sector can create significant numbers of productive jobs (as in jobs that produce goods or services that can be profitably sold, swelling government tax coffers to enable investment in social infrastructure like libraries, concert halls, convention centres, schools, hospitals, the blue light services (police, fire, ambulance) roads, metros and railways). Mason (2010) ventures: '[T]he American public sector seems very poorly geared to spending money.' Another interpretation is that better results could have been achieved had some (or all?) of the Obama administration's fiscal stimulus money been channelled through community groups and small business associations instead of state and city authorities. Theoretically the Cameron–Clegg Big Society project provides a vehicle for bottom-up, community-directed economic growth (Bennett 2010, Cameron 2010). The organic approach is not without its problems, however, like standards and accountability. Before the National Health Service, standards of health provision were more variable. Community groups are not as democratic as, say, town councils whose members are periodically judged by the electorate in a secret ballot.

8   'Some forty square miles has evolved past decrepitude into vacancy and prairie' (Solnit 2007).

vortex of social and economic decline saw a budget shortfall of nearly $300 million (Behunek 2010). The city issued municipal bonds, and hoped.

Detroit's troubles have a long genesis. During the Second World War, despite a huge increase in production and opportunity, there were racial tensions and riots. In July 1943 two nights of violence saw 34 people killed. The government put troops on the street. During the war it confronted racial intolerance abroad *and* at home.

Between 1940 and 1960 'the proportion of blacks in the population grew to one-third. The white middle-classes, full of resentment against the black lower-classes, fled to the periphery [the so-called "white flight"]' (Mende and Oswalt 2004). In 1967 after years of discrimination, exploitation and police harassment the city's African-American population took to the streets. The riots, a product of institutionalised racism,[9] subjugation and alienation, were a watershed event (Feldman 2010). The city's bifurcation on racial lines accelerated. More and more whites moved to the suburbs: 'There were 1,600,000 whites in Detroit after the war, and 1,400,000 of them left. By 1990 ... the suburbs had surpassed Detroit not only in population, but in wealth, in commerce ...' (Young cited in Solnit 2007). As the city's legitimate economy shrank, the drug economy grew. Kerr (2000) observes:

> *Although segregation was already a fact of life ... the riots left an enduring legacy of extreme polarisation that further contributed to the community's social collapse. By the 1970s Detroit had acquired the thoroughly deserved nickname of 'Murder City'. Its 1974 record of 714 homicides was the worst in the nation, a feat it matched in several other years in the following decade.*

Determined to get his city back on the rails, Detroit's first African-American Mayor, Coleman Young, forged links with the city's mainly white business community. When Young's enthusiasm for corporatist, inclusive politics waned in the 1980s the city's business community fractured along racial lines. While significant investments were made (for example, in General Motors' new Poletown plant and Chrysler's new Jefferson Avenue plant) Detroit's politicians seemed more interested in holding on to power than in their city's economic and social decline:

---

9   Solnit (2007) remarks: 'Detroit was ... a segregated city with a violently racist police department and a lot of white people ready to work hard to keep black people out of their neighbourhoods.'

> [T]he corporatist alliance between the white downtown business
> establishment leaders and the Young administration fell victim to racial
> and clientelistic politics. (DiGaetano and Lawless 1999)

Because, under the US system of governance, so much power is vested in
local authorities (DiGaetano and Lawless 1999), when that system fails (as
in Detroit in the 1980s) there is little hope for the locale. Detroit's fall from
grace continued. While the election of a new city regime in the 1990s helped re-
engage the business community, the damage had been done. The city region's
de facto apartheid engendered feelings of alienation and hopelessness amongst
inner-city blacks:

> The growing divide between Detroit and its suburbs compounded the
> lack of opportunity. [Coleman] Young argues that isolation and racism
> by the suburbs led to sentiments of self-hate and self-destruction among
> the city's residents. (Moceri 2003)

According to the US Census Bureau, by the autumn of 2005 Detroit had become
the poorest city in America. Those who could leave had gone. Those who could
not hunkered down:

> Ever since 2000 there has been a continuing exodus from the city.
> Anybody with the wherewithal, especially those with children, is finding
> ways to get out … Those who can't afford it aren't going anywhere.
> So there is no reason why the poverty rate in Detroit shouldn't be so
> high. There are pockets of gentrification, but that's a drop in the bucket.
> The city has so many people without education, without skills, without
> anything. Nobody is creating jobs, they can't go anywhere … I would
> not be surprised [if the real unemployment rate is] closer to 30 or 35
> percent. (Metzger cited in Walsh 2005)

In 2010 the British Broadcasting Corporation reported from Detroit:

> Drive through Detroit and immediately you see the scale of this city's
> problems. There are burnt out houses, piles of rubbish and empty lots
> on every block. Anyone who can seems to have fled. The city was built
> for two million but now has a population of only 800,000. So 40,000
> acres of Detroit now stand unused, home to weeds, broken glass, even
> pheasants. (Kay 2010)

Motor City – Modernity Writ Large?

Before its present difficulties, Detroit was a crucible of progress – a shining beacon of modernity in a new land, untroubled by tradition. Henry Ford set the tone for Detroit's development when he intoned:

> *History is more or less bunk. It's tradition. We don't want tradition. We want to live in the present, and the only history that is worth a tinker's damn is the history we make today. (Ford cited in Kerr 2000)*

Ford and his contemporaries were unsentimental modernisers (Murray 1989, Hoffman 2003). They were rationalists. To sell more cars Ford knew he had to reduce his production costs. Ford reified his economic ambition in the moving assembly line (introduced first at his Highland Park plant in Detroit). The innovation worked – the purchase price of his Model T fell from $850 to $298. To enable the public to mass consume his mass produced automobiles Ford introduced the $5-a-day wage. Another success – his Model Ts flew out of the showrooms. Scientific management allowed Ford to reduce the time it took to make a Model T from 12.5 hours to 93 minutes.[10] Ford's modern approach was infectious. Soon other manufacturers began building expansive, steel-reinforced, well-lit factories to house their moving assembly lines. Despite several setbacks, like the 1929 Wall Street crash and subsequent deadly union recognition disputes, by the end of the 1930s Detroit was an industrial powerhouse. Vast factories employed hundreds of thousands of workers. New highways were built to speed traffic to and from Detroit's increasingly salubrious city centre. Detroit was a *modern* metropolis built on rational principles (albeit with the occasional baroque architectural flourish):

> *With a war looming, Detroit could look back on an era of unprecedented growth and prosperity. Not only had it become the forge of America's new industrial empire; the city itself had taken on the guise of a great metropolis. High-rise offices and hotels had sprouted in the downtown area ... By contrast, the enormous automotive factories ringing the city had become celebrated icons of the European Modern movement. For just as the Detroit industrialist Henry Ford had revolutionised the means of production, so the Detroit architect Albert Khan had pioneered a new form of architecture in which to house the new assembly lines ... [Khan's designs] provided large, uninterrupted spaces between columns, essential*

---

10 On a single day in 1925 Highland Park produced 9,000 automobiles. At its peak Ford's River Rouge plant employed 100,000 Detroiters.

for the new production methods, and allowed natural light to flood in from floor-to-ceiling windows. Detroit became an essential pilgrimage destination for avant-garde artists and architects, eager to witness the miracle of mass production. Images of Kahn's Highland Park factory appeared in all of the pioneering Modernist texts ... The Rouge plant [was] universally received as the epitome of modernity ... (Kerr 2000)

Detroit's social, economic and political life was dominated by a singular discourse – that such instruments of modernity as rationalism, optimisation, innovation in social action, economy of effort, scientific management (Taylorism), mass production (Fordism), mass consumption, mobile international capital, free market economics and labour market flexibility could deliver the best possible outcome for Detroit and its burgeoning population. Innovations – many of them connected to the motor industry – came thick and fast. In 1901 Detroiters benefited from the world's first concrete road. In 1915 Detroit commissioned the world's first traffic light. In 1954 Detroiters were treated to the world's first shopping mall – modern, modernist, and reliant upon a car-borne clientele (Herron 2004). The centrifugal pull of the suburbs continued.

Detroit's rejection of sentimental attachment and willingness to remake itself provided fertile ground for all things modern: 'Monuments have short half-lives in [Detroit]. Things move too quickly and survival is more a matter of forgetting than remembering' (Hoffman 2003). In his essay 'Fordism and Post-Fordism' Murray (1989) associates Fordism with modernism:

[Fordism] is marked by its commitment to scale and the standard product ... by a competitive strategy based on cost reduction; by authoritarian relations, centralised planning, and a rigid organisation ... The technological hubris of this outlook, its Faustian bargain of dictatorship in production in exchange for mass consumption, and above all its destructiveness in the name of progress ... all this places Fordism at the centre of modernism.

While the Fordist discourse was not universally accepted or liked (as evidenced by the labour unrest of the early 1930s sparked, in part, by the halving of the hourly pay rate from 92 to 52 cents) it dominated city politics for much of the twentieth century. Certainly there were setbacks. Between 1945 and 1965 Detroit lost 150,000 manufacturing jobs. That symbol of progress, the urban motorway, scythed through communities leaving them to atrophy: 'By the mid-1950s, 7,000 Detroiters had been displaced by motorways, and many previously flourishing inner-city neighbourhoods had been bisected by them, precipitating rapid social decline' (Kerr 2000).

Despite its negative side effects, Detroit's political rulers never lost faith in the modernist project. The received wisdom was that Detroit's future lay in the hands of the major corporations (subsidised by the municipality, of course) and technocratic elites. Following Detroit's social nadir of the 1967 riots its politicians looked to the Big Three motor companies for solutions. During the 1970s Detroit's first black mayor courted Detroit's white business elite, building a social and political consensus around major corporate investment. The answers to Detroit's problems were seen to lie with corporate decision-makers:

> *In the early 1970s the city's business establishment took the initiative to advance a progrowth policy agenda ... [T]he city's downtown business leaders formed Detroit Renaissance, which became the organisational muscle of Detroit's white corporate establishment on economic development matters ... Coleman Young ... adopted a surprisingly conciliatory relationship with the Motor City's white business establishment. Indeed, a corporatist governing structure was forged around the city's downtown renaissance agenda. (DiGaetano and Lawless 1999)*

Corporate and political energies and monies were channelled through three development corporations, whose governing boards were 'composed primarily of corporate executives and public officials' (DiGaetano and Lawless 1999). Major developments were heavily subsidised by the city. General Motors' Poletown plant received $200 million in public subsidies, while Chrysler's Jefferson Avenue plant received $264 million.

Detroit's response to its post-1973 oil crisis decline referenced the traditional approach to social and economic development – retain existing manufacturing, attract new manufacturing, encourage the service and entertainment sectors, provide employers with development land, modern infrastructure and a skilled workforce and, where necessary, subsidise new developments. This approach had worked in the past. Why should it not work in the future? There were some successes. The Big Three invested in new manufacturing facilities and the city spawned a new sector – gambling. By 2003 three casinos had been opened in downtown Detroit. But Detroit was not Las Vegas:

> *Aware of the perceived danger of its downtown location, the Motor City Casino publicity boldly states, 'The area is not pretty, absolutely no local ambience. That's alright because all the action is inside'. (Hoffman 2003)*

Detroit's orthodox top-down, technocratic approach failed to stop the rot, however. White flight continued, as did business migration, often to states or countries with lower labour costs. Cope (2004) talks of 'a perpetual motion outwards ... in an eternally increasing pace'.

Technocracy seemed to be failing Detroit's increasingly desperate citizens. Risks multiplied and grew. Real incomes fell. Homes and businesses were abandoned. A lower tax take saw taxes rise. The schools system began to fail. A 'crack economy', fuelled by desperation and underwritten by gang violence, emerged (Hoffman 2003). Detroit became known as America's Murder Capital. Arson rates soared.[11] Motor City seemed about to self-destruct:

> *The increase in arsons [in 1981 there were 6,204 arsons and 8,607 fires of suspicious origin] and their cost was partially due to an increase in fraud. Detroit had increasing numbers of absentee property owners who found burning their properties more profitable than maintaining them. (Moceri 2003)*

Disaffected citizens burned empty properties during Halloween. During the 1984 three-day Halloween period Detroit suffered 810 fires. Detroit's former Fire Chief recalls:

> *1984 was our worst 'Devil's Night' [Halloween], the worst fire scenes I have seen since the riots of 1967. We had fires burning where there were no fire companies available to respond .... It was the worst thing I have seen on a non-riot basis. (Bozich cited in Moceri 2003)*

There were pressures within, and pressures without. The Federal government withdrew much-needed aid (during the 1980s Federal assistance to Detroit was cut by almost 50 per cent (DiGaetano and Lawless 1999)):

> *The Reagan presidency reversed a half-century of federal aid to cities. Poor minority communities were particularly hard-hit, since this was accompanied by a white flight to the suburbs and the replacement of better-paying industrial jobs requiring little education with poorer-paying service jobs requiring more education. (Stoesz 1992)*

The American public bought more imports – Japanese and German cars were better engineered and better built (Kerr 2000). A globalising world economy sucked jobs

---

11 According to Solnit (2007) some of the arson was 'constructive' in that it razed drug dens.

out of the United States, hitting America's rust-belt cities like Detroit especially hard. Capital and jobs migrated to low-cost economies.[12] For Bauman (1998) the modern multinational corporation (MNC) represents 'unanchored power':

> *It is up to [shareholders] to move the company wherever they spy out or anticipate a chance of higher dividends, leaving to all others – locally bound as they are – the task of wound-licking, damage-repair and waste-disposal. The company is free to move; but the consequences of the move are bound to stay. Whoever is free to run away from the locality, is free to run away from the consequences. These are the most important spoils of victorious space war.*

In Detroit the consequences of numerous moves were abandoned buildings, waste capital, derelict land and social dysfunction. During the 1980s these impacts were compounded by the politics of race (DiGaetano and Lawless 1999). The city's decline accelerated during the 1980s. Detroit never recovered. Costly interventions failed to stop the rot:

> *Over the decades, grand schemes to renovate Detroit have come and gone, leaving almost no improvement. There have been shopping centres, sports stadiums and even casinos. None have made much impact on Detroit's 28% unemployment rate. (Kay 2010)*

Moon (2009) notes:

> *[Detroit's] grand gestures – casinos, ballparks, and corporate headquarters – have not only failed to resuscitate the city; they stand out as monuments to disappointment. What the city needs is hope – hope that a solution to current crises will emerge and that it will transcend the myth of salvation.*

Depopulation and Disinvestment – the Experience of the Parish of St. Cyril, Detroit

These two photographs (see Figure 9.1), taken 54 years apart, bear witness to urban Detroit's dilapidation, depopulation and spontaneous greening. This

---

12 Some Republican politicians blame high-taxing civic authorities for industrial migration. Indiana's governor says: '[P]eople aren't leaving here because you didn't tax them enough ... they left because you taxed them too much or you simply did not create the conditions for a private sector to flourish' (Daniels cited in Mason 2010).

Figure 9.1   Depopulation and disinvestment – The experience of the Parish of St. Cyril, Detroit

pattern is typical of the inner city. The area's transformation began in the 1970s with the closure of the elementary and high school (in 1971). The loss of social infrastructure continued into the 1980s with the relocation of the church of St. Cyril to the suburbs. The last service at St. Cyril was held in December 1988. By 2000 all of the parish buildings had been abandoned. Anything of value was quickly removed.

Detroit is not alone in experiencing disinvestment and depopulation. According to the Brookings Institution the trend afflicts 50 US cities, including Memphis, Baltimore, Pittsburgh, Philadelphia, Flint and Detroit. Daniel Kildee,

**Figure 9.2   Apogee: Detroit in 1942. Untrammelled power and optimism.
The underpinning of the Free World**

the author of the 'shrink to survive' policy for North America's declining rust-belt cities, highlights America's 'expansionist' mind-set as an obstacle to urban restructuring and down-sizing: 'The obsession with growth is sadly a very American thing. Across the US there is an assumption that all development is good, that if communities are growing they are successful. If they're shrinking, they're failing' (Kildee cited in Leonard 2009). The Director of the Shrinking Cities in a Global Perspective programme at Berkley claims that many are in denial about US decline. Denialism frustrates reform.

8 MILE LESSON?[13]

In his essay *Shrinking City Detroit*, Park (2004) asks whether de-urbanisation could be turned to the advantage of those who are left behind. Orthodoxy holds that de-urbanisation is a social bad. Park challenges this view:

---

13 8 Mile Rd (actually a multi-lane highway) runs west–east eight miles from the centre of Detroit. To some it divides the impoverished inner city from the more prosperous (and mainly white) northern suburbs. The road is fronted by low-end convenience stores, strip clubs, furniture rental and cheque-cashing shops. The road is an allegory of Motor City, a miasma of hedonism and poverty, functionality and disamenity, speed and creeping decay. As one subscriber to urbandictionary.com puts it: 'There are blacks, whites and trailer trash all along it'. The further north you venture, the more salubrious the surroundings. Eminem, star of the film 8 Mile, was born in Warren, a lower-middle class, mostly white city to the north of 8 Mile Rd. The film charted the rapper's struggle for acceptance amongst Detroit's black musicians. It also showed what it is like to work long hours for little money in a disinvested rust-belt industry (metal pressing).

> *Is immediate demolition or development really necessary, or can cultures and cities live with ruins? ... Must ... empty buildings and spaces be moulded to the existing pattern of use and to existing urban practices? ... Can there be mutually beneficial relations between unoccupied buildings and an unemployed population? Could these empty spaces be given for free to the unemployed population to self-incubate new types of labour and economy in informal, separate or semi-autonomous states?*

Tired of waiting on the authorities, some Detroiters have taken matters into their own hands. Detroit's anarchic urban agriculturalists have planted vacant land and make money selling produce to a population hungry for locally-grown and organic food:

> *[In Detroit] there is open land, fertile soil, ample water, willing labour, and a desperate demand for decent food. And there is plenty of community will behind the idea of turning the capital of American industry into an agrarian paradise. In fact, of all the cities in the world, Detroit may be best positioned to become the world's first one hundred percent food self-sufficient city. (Dowie cited in Renn 2009)*

Other Detroiters have established charitable foundations or have proposed innovative new industries. Such initiatives vary in the degree to which they articulate late-modern ideals (subversion, diversification, self-reliance, inclusivity, unconventionality and empowerment of the grass-roots, for example). The spontaneous planting of vacant land by unemployed residents and incomers represents the purest articulation of the late-modern blueprint. A financial tycoon's proposal to establish an agri-business in downtown Detroit the least. Detroit's many charities and neighbourhood associations sit somewhere in the middle.

Alternative Voices

On 13 March 2010 the BBC broadcast *Requiem for Detroit*, a portrait of the Motor City past and present. Several of Detroit's growing cadre of small-scale urban agriculturalists were interviewed around a fire on a piece of ground. Unbothered by the authorities, contented and with cash in their pockets, they spoke enthusiastically of their agrarian colonisation of the city:

> Settler 1: The amazing thing to me about Detroit is that it brings people in who are risk-takers, who can implement things they believe in.

Settler 2: You come here for a few days and you are, like, 'wait a second, this place is amazing'. On paper it is the worst place on earth, you know, and it is the last place you ever want to go to.

Settler 3: What happened to Detroit? It was a man-made disaster, not a natural disaster. I suddenly realised that we were going to let a major American city just fall off the face of the earth. I thought 'I want to go into the belly of the beast to see if I can figure out what is going on and whether we can make things better'.

Settler 1: Fortunately our city government isn't actively keeping track of what is going on. So a lot of us are growing our own food. I want to test the theory that you can make a decent living on an acre of land. Right now it is bringing in about $500 a week.

Resident 1: Which is about what I make at Chrysler in a five day week. My people come from down south. We still have 40 acres down there that we used to farm, so it is not really anything new to me. Actually it is better now than when I thought I was really doing good. How could I live like this? How could I not live like this?

Settler 1: This developing cottage industry is pretty much going to be Detroit's economy in ten years, instead of regular industry.

Resident 1: I don't think it is going to be that long.

Grace Lee-Boggs, author, activist and native of Detroit, was also interviewed:

> *At least 50,000 houses have been bulldozed. When the lots became vacant many older African Americans who had been raised in the south and who had grown their own food, decided to plant community gardens, not just for the food, but because they realised that food is the way that you begin to care for yourself and think about yourself in a different way. The fastest growing movement in the United States today is the urban agriculture movement. What has been so exciting is the number of young people who come to Detroit because they feel that we are pioneers for the 21st century. The old American dream is dead. We are in the process of creating a new American dream. It is happening in Detroit – city of hope. You can look at Detroit and see nothing but disaster, devastation, depopulation, disinvestment. Or you can look at Detroit and say 'That is the future'.*

In her essay *Detroit's 'Quiet Revolution'* Boggs (2009) describes how Detroit's nascent urban agriculture movement confronted Mayor Young's administration when it proposed a gambling industry as part of the solution to Detroit's problems. Boggs helped set up a lobby group – Detroit Summer – to launch youth projects, community gardens and other grass-roots initiatives. A spin-out from Detroit Summer, the Detroit Agricultural Network, 'includes more than 700 community gardens, lovingly cultivated by Detroiters of all ages, walks of life and ethnicities, who share information on resources, and how to preserve and market produce ...' (Boggs 2009). As far as Boggs (2009) is concerned, Detroit's political class is 'Stuck in the old social democratic paradigm, in which government stimulus programmes ... are seen as the only methods of reviving communities ... '. For Boggs, the best solutions draw on local expertise and labour. They percolate up from the street:

> [T]hose in control are dysfunctional ... the old social democracy dependence on those in power to give you things, that period is over ... We are creating a new culture ... We're doing it because we had to ... I think that anyone who attempts a top-down solution can't succeed ... The answers can't come from the top ... The answers are coming more from the bottom (Boggs cited in Goodman 2010b).

The Chair of the Detroit Black Community Food Security Network claims that politicians cannot and should not be trusted:

> What I hope to see is a ramped-up or an increased urban agriculture movement in the city of Detroit. I hope to see more understanding of the value of this movement by the political leadership within the city ... [B]ut what's most important is the people being mobilized and the people's consciousness being raised, so that they see this as being a way to work on their own behalf. Once people are mobilized and see that they can begin to work for their own benefit, then that's what holds politicians accountable. If people aren't mobilized and able to speak on their own behalf and act on their own behalf, then politicians come in and do whatever they want to do. (Yakini cited in Goodman 2010a)[14]

---

14 It could be argued that Detroit's urban farmers with their small-scale strip cultivation represent a pre-modern social form (strip farming was a characteristic of Feudal societies). Consequently, Detroit could be said to exhibit late-modern, modern and pre-modern social formations. This is very much the pattern observed in newly industrialising countries like China and Brazil. In urban Detroit is 'regression' progression?

The ground for Detroit's urban agriculture movement was laid in the late 1980s by, amongst others, union organiser James Boggs. Boggs dished capital and promoted self-reliance:

> We have to get rid of the myth that there is something sacred about large-scale production ... We have to begin thinking of creating small enterprises which produce food, goods and services for the local market. (Boggs cited in Solnit 2007)

Perhaps surprisingly, the UK Conservative Party – in coalition with the Liberal Democrats – espouses a similar Big Society thesis. It is illuminating to compare David Cameron's Big Society vision with that of radical activist Grace Lee-Boggs:

> [L]et me briefly explain what the Big Society is ... The Big Society is about a huge culture change, where people, in their everyday lives, in their homes, in their neighbourhoods, in their workplace don't always turn to officials, local authorities or central government for answers to the problems they face, but instead feel both free and powerful enough to help themselves and their own communities ... For years, there was the basic assumption at the heart of government that the way to improve things in society was to micromanage from the centre, from Westminster. But this just doesn't work. We've got the biggest budget deficit in the G20. And over the past decade, many of our most pressing social problems got worse, not better. It's time for something different, something bold – something that doesn't just pour money down the throat of wasteful, top-down government schemes. (Cameron 2010)

While Cameron's plan differs in some important respects from that of Boggs and her associates (in respect of, for example, the setting up by government of a Big Society Bank, and funding of Community Organisers for each locale) the spirit is very much that of a late-modern shift of power from politicians and technocrats to citizens and grass-roots organisations. The expectation is that ideas will filter up from the street. After all, who knows a community best? Whitehall civil servants, Town Hall functionaries, or residents? Which group has most to lose if projects fail? Who experiences the consequences of failure most acutely?

While there are potential benefits from such a divestment of power, there are also potential pitfalls. How, for example, do central and local governments ensure that the same standards are achieved across the country?[15] Is it not possible that prosperous locales will draw ahead of poor locales in terms of community-actioned schemes? After all, money buys leisure and freedom. The middle classes, by virtue of their greater prosperity, are more able to invest time and energy in community action. A more sinister development would be the appropriation of grass-roots initiatives by big business. Even as late-modern social phenomena (like urban agriculture and small, independently-owned community workshops) become more evident, capitalistic enterprise still determines the economic and social (and to some degree, political) agenda. Citizens may be mobilising, but they are no match for global capital. Multinationals are powerful and discriminating and function largely without sentiment (Bauman 1998).

The Appropriation of Late-Modern Social Formations

Like Grace Lee-Boggs, John Hantz is a native of Detroit. Unlike Boggs, Hantz is a multi-millionaire. With a personal net worth of around $100 million Hantz plans to create the world's largest urban farm right in the heart of Detroit. Although motivated by a desire to rebuild the shattered hulk that is downtown Detroit, Hantz's farm is about as far removed from Lee-Boggs's vision of civic agrarianism as it is possible to get. Hantz plans 'a high-tech farm, growing a mixture of vegetables, fruit and trees using the latest agricultural science ...' (Kay 2010). Initially Hantz planned a single green swath that would be added to year by year. After talking to the Kellogg Foundation, however, he was persuaded to acquire geographically dispersed 300-acre plots of land. These plots, which Hantz called 'lakes', would generate income for Hantz, jobs for residents and stimulate the city's moribund property market by creating an extensive green frontage – something pleasant for tenants and owners to overlook. Desirable farm-fronting properties would sell, stimulating the market. As more and more people drifted back to Detroit the tax take would rise. The city's economy would start to expand, albeit in a different direction. City Hall would have more to spend on Detroit's failing schools system and fractured infrastructure. For Hantz, the beauty of his project is that it creates scarcity in a property market stultified by abundance. Whitford (2009) explains Hantz's real estate rationale:

> *Detroit looks pretty good right now to a young artist or entrepreneur who can't afford anyplace else – but not yet to an investor. The smart*

15　See the earlier comments on the plight of Gary, Indiana.

> *money sees no point in buying as long as fresh inventory keeps flooding the market. 'In the target sites we have', says Hantz, 'we [re-evaluate] every two weeks'. As Hantz began thinking about ways to absorb some of that inventory, what he imagined, he says, was a glacier: one broad, continuous swath of farmland, growing acre by acre, year by year, until it had overrun enough territory to raise the scarcity alarm and impel other investors to act.*

Hantz's lakes would use efficient high-intensity, high-technology farming methods – agricultural Fordism in the city that pioneered mass production (of motor cars, trucks, tanks, aero engines, munitions, etc.):

> *In fact, Hantz's operation will bear little resemblance to a traditional farm. Mike Score, who recently left Michigan State's agricultural extension program to join Hantz Farms as president, has written a business plan that calls for the deployment of the latest in farm technology, from compost-heated greenhouses to hydroponic (water only, no soil) and aeroponic (air only) growing systems designed to maximize productivity in cramped settings. (Whitford 2009)*

Detroit's political and business establishment have welcomed Hantz's idea, if only because it capitalises on Detroit's relentless hollowing out. Whitford (2009) explains:

> *Nearly 2 million people used to live in Detroit. Fewer than 900,000 remain. Even if, unlikely as it seems, the auto industry were to rebound dramatically and the US economy were to come roaring back tomorrow, no one – not even the proudest civic boosters – imagines that the worst is over. 'Detroit will probably be a city of 700,000 people when it's all said and done', says Doug Rothwell, CEO of Business Leaders for Michigan. 'The big challenge is, What do you do with a population of 700,000 in a geography that can accommodate three times that much?' … [T]here's the problem of what to do with the city's enormous amount of abandoned land, conservatively estimated at 40 square miles in a sprawling metropolis whose 139-square-mile footprint is easily bigger than San Francisco, Boston and Manhattan combined.*

Speaking to the concept of a well-funded, high-technology urban agri-business, The Chief Executive Officer of the Detroit Economic Growth Corporation (DEGC) has said:

We have to be realistic. This is not about trying to re-create something.
We're not a world-class city. [Hantz's idea] sounds very exciting. We
hope it works. (Jackson cited in Whitford 2009)

The DEGC has incentivised the project by offering free land. Detroit's Mayor
has lent his support:

[I am] encouraged by the proposals to bring commercial farming back to
Detroit. As we look to diversify our economy, commercial farming has
some real potential for job growth and rebuilding our tax base. (Bing
cited in Whitford 2009)

The American Institute of Architects (AIA) has welcomed the initiative, saying:
'Detroit is particularly well suited to become a pioneer in urban agriculture at
a commercial scale' (American Institute of Architects cited in Whitford 2009).[16]
Endorsements like these legitimise Hantz's top-down, business-centric approach.

We should not be surprised at the support Hantz has received from traditional
authority figures and networks. Hantz's lakes would generate tax dollars and
pump wage income into the city economy. His use of high-technology growing
systems (like hydroponics) would spur innovation in America's green/sustainable
technology industry – a sector promoted by the Obama administration. Detroit
City Hall still has a vast hard and soft infrastructure to maintain (sewage
systems, schools, garbage collection, law-enforcement, fire-fighting, etc.). For
all their pioneering verve, those who till the 900 or so non-profit plots overseen
by the Detroit Agriculture Network (DAN) contribute little in monetary terms
to City Hall. Detroit's politicians are looking for big solutions to big problems.
The Fordist production methods that underwrite Hantz's scheme are familiar.
Hantz's concept of a well-funded, large-scale, high-technology, (potentially)
high-return, 'legitimate' agribusiness resonates with the traditionalists. Fordism
once made Detroit great. Why not again?

Those opposed to Hantz's scheme highlight the fact that it would employ
relatively few people (cost-effectiveness and optimisation are characteristics of
Fordism) and would not be owned by the community. With reference to Marx's
theory of alienation, the product of workers' labour would be appropriated by

---

16 Paradoxically, perhaps, the AIA also endorses Detroit's small-scale urban agronomists. It has
called for: 'Increase[d] support for existing community gardens and urban agriculture in the
city, including focused support for increased productivity and expansion of Detroit's existing
commercially – or community – oriented urban farms' (American Institute of Architects 2008).

the owner, who would make all the key decisions. Workers would be 'wage slaves'. The Chair of the Detroit Black Community Food Security Network comments: 'I'm concerned about the corporate takeover of the urban agriculture movement in Detroit' (Malik cited in Whitford 2009).

Faced with such opposition Hantz has gone on the offensive. 'Someone must pay taxes' insists Hantz (cited in Whitford 2009). For Hantz, the choice facing City Hall is between a tax-paying, job-creating, tourist-attracting agribusiness and an unregulated, unpredictable, non-tax paying social movement:

> *Detroit's fire, police and public works departments can better serve city residents when freed from the burden of nearly abandoned neighbourhoods. (Hantz Farms 2009)*

Where does Detroit's Future Lie?

Detroit's future is being shaped in the midst of an ideological contest. On the one side there are Detroit's urban pioneers who invest their time and energy in small-scale, not-for-profit urban agriculture schemes and independent workshops. Detroit's counter-culture is as innovative and entrepreneurial as any major corporation – perhaps more so. This irony should not be lost on General Motors, Ford, Chrysler, entrepreneur John Hantz and Mayor Bing. America owes everything to its early pioneers – those intrepid, self-reliant adventurers who beat a path westwards to the Pacific. Today Detroit is indebted to those who are re-colonising Motor City's downtown wastelands. These long-time residents and enthusiastic incomers are America's twenty-first century pioneers.

On the other side are those like Hantz (encouraged by Mayor Bing, the DEGC and AIA) who advocate technocratic, capitalistic, high-technology solutions to Detroit's chronic social and economic problems. Fordism, a philosophy of production familiar to those in power, is still influential. The mass-production paradigm persists. Given this fact, what chance do Detroit's counter-culturalists have? As happened to San Francisco's counter-culture movement in the 1960s, will Detroit's new millennium counter-culture movement be mocked, trivialised, undermined, enrolled and eventually consumed by big business? Can such late-modern social formations as collectives and small independent producers withstand the ambitions and appetites of big business? Can non-conformity survive? Are we becoming more, or less tolerant of multiple world-views – of multiple realities? Is there a future for difference? If not, what does this say about the (still fashionable) theory of late-modernity?[17] What does it say about concepts like 'technological citizenship' (Irwin 1995) and 'technological governance' (Zimmerman 1995) which support wider participation in decision-making?

In 2010, City Hall, with financial aid and encouragement from Washington, launched another top-down initiative:

> The ... plan would demolish about 10,000 houses and empty buildings in three years and pump new investment into stronger neighbourhoods. In the neighbourhoods that would be cleared the city would offer to relocate residents or buy them out. The city could use tax foreclosure to claim abandoned property and invoke eminent domain for those who refuse to leave, much as cities now do for freeway projects. (Huffington Post 2010)

While the aid package of $40 million was welcomed by Mayor Bing and the Detroit Housing Commission, others were sceptical:

> Maggie DeSantis, a board member of Community Development Advocates of Detroit, said she worries that shutting down neighbourhoods without having new uses ready is a 'recipe for disaster' that will invite crime and illegal dumping. (Huffington Post 2010)

Given that there is little chance of attracting new development in the short or medium term, clearance would have the effect of creating a physical and social vacuum. Moon (2009) asks: 'In Detroit, erasure has led to a new crisis of emptiness ... Why create voids without plans for recuperation?' Clearance would also create the impression that those in authority have the answers to Detroit's malaise, when, as recent history shows, they do not. Detroit's urban agriculturalists, on the other hand, have tackled the problem of abandonment head-on, and with some success. Land has been put to productive use in days rather than months or years.

Urban agriculture adds value. It provides employment, albeit on a small scale. Most importantly, it creates hope. In creating a space for self-expression it builds confidence and self-respect. Atomised communities begin to cohere. A brutalised cityscape begins to heal. Moon (2009) identifies three distinct phases in 'bottom-up' reclamation. In the first phase, problems are highlighted

---

17 A late-modern society is 'a more fragmented and variegated society' (Hall and Jacques 1989).

by grass-roots activists. In the second, resources (land, buildings, etc.) are appropriated. In the third, abandoned lots and buildings are transformed into viable entities. For Moon (2009) the answer to Detroit's problems lies in spontaneous civic action, in new suburbanism:

> Detroit may be littered with ruins and abandoned lots, but it is not empty. In response to architectural neglect and decay, individual acts of appropriation suggest that the remedy for post-industrial ruin begins with ad hoc, underground, and unsanctioned practices. By calling attention to derelict and downtrodden conditions, Detroit residents are, in a variety of ways, initiating an urban revival. Resuscitating Detroit will require a multilayered strategy, one that is currently bringing signs of revitalisation, but whose future is uncertain.

The big question, of course, is whether the authorities are willing to facilitate such grass-roots initiatives as Detroit's urban agronomists. Bureaucracies are ordered. Bureaucracies seek to order. They plan for the future. They devise policy. They set performance targets. They seek feedback. They (generally) act within the law. Detroit's citizen-planters are anarchic, messy, spontaneous, unfathomable, uncoordinated and unpredictable. The question, therefore, is whether Detroit's and Washington's technocrats are willing to tolerate the city's non-conformists – its free-form agronomists, squatters and small-scale entrepreneurs? Can order co-exist with subversion? Are the tensions containable? Can late-modern social formations (like communes or independent strip cultivators) co-exist with big business (in Hantz's case, agribusiness) and big government? Or are they destined to be either appropriated or annihilated? Orwell (1950) wrote:

> If you want a picture of the future, imagine a boot stamping on a human face – forever.

Was he right?

References

American Institute of Architects. 2008. *Leaner, Greener Detroit: A Report by the American Institute of Architects Sustainable Design Assessment Team*. Washington, DC: American Institute of Architects.

Andersen-Brower, K. and Johnston, N. 2010. Obama says auto industry 'growing stronger', Creating Jobs. *Bloomberg.com*, 30 July. Available at: http://www.bloomberg.com [accessed: 27 August 2010].

Bauman, Z. 1998. *Globalisation – the Human Consequences*. Cambridge: Polity.

Beck, U. 1992. *Risk Society: Towards a New Modernity*. London: Sage.

Beck, U. 2009. *World at Risk*. Cambridge: Polity.

Behunek, S. 2010. Three American cities on the brink of broke. *CNNMoney.com*. Available at: http://money.cnn.com [accessed: 22 July 2010].

Bennett, R. 2010. The Big Society has made little impact, says think-tank. *The Times*, 6 September.

Billups, A. 2009. Detroit haunted by poverty, scandal. *The Washington Times*, 22 February.

British Broadcasting Corporation. 2009. Requiem for Detroit. London: Films of Record. Broadcast 21:00–22:15, 13 March 2010, BBC Channel 2.

Bruce, S. 1999. *Sociology: A Very Short Introduction*. Oxford: Oxford University Press.

Cameron, D. 2010. Big Society speech. Available at: http://www.number10.gov.uk/news [accessed: 22 July 2010].

Cope, M. 2004. In Detroit time, in *Working Papers: Detroit*. March, 2004, edited by D. Mende and P. Oswalt. Available at: http://www.shrinkingcities.com [accessed: 1 July 2010].

Davis, J. 2010. Return to the Red Planet. *Telegraph Magazine*, 16 October.

DiGaetano, A. and Lawless, P. 1999. Urban governance and industrial decline: Governing structures and policy agendas in Birmingham and Sheffield, England, and Detroit, Michigan, 1980–1997. *Urban Affairs Review*, 34(4), 546–77.

Feldman, R. 2010. The rise and fall of the auto industry in Detroit. *Democracy Now: The War and Peace Report*. Available at: http://www.democracynow.org [accessed: 19 July 2010].

Giddens, A. 1999. Risk and responsibility. *Modern Law Review*, 62(1), 1–10.

Goodman, A. 2010a. Detroit Urban Agriculture Movement Looks to Reclaim Motor City. *Democracy Now: The War and Peace Report*. Available at: http://www.democracynow.org [accessed: 19 August 2010].

Goodman, A. 2010b. 'The Answers Are Coming from the Bottom': Legendary Detroit Activist Grace Lee-Boggs on the U.S. Social Forum and her 95th Birthday. *Democracy Now: The War and Peace Report*. Available at: http://www.democracynow.org [accessed: 19 July 2010].

Gorz, A. 2000. *Reclaiming Work: Beyond the Wage-Based Society*. Cambridge: Polity.

Groom, B. 2010. Britain: Balance and power. *Financial Times.com*, 21 July. Available at: http://www.ft.com [accessed: 27 August 2010].

Hall, S. and Jacques, M. 1989. Introduction, in *New Times – The Changing Face of Politics in the 1990s*, edited by S. Hall and M. Jacques. London: Lawrence and Wishart, 11–20.

Hantz Farms. 2009. Introducing Hantz Farms. Available at: http://www.hantzfarmsdetroit.com [accessed: 29 August 2010].

Held, D. 1989. The decline of the nation state, in *New Times – The Changing Face of Politics in the 1990s*, edited by S. Hall and M. Jacques. London: Lawrence and Wishart, 191–204.

Herron, J. 2004. Chronology: Detroit since 1700, in *Working Papers: Detroit*. March, 2004, edited by D. Mende and P. Oswalt. Available at: http://www.shrinkingcities.com [accessed: 1 July 2010].

Hoffman, D. 2003. The capital of the twentieth century, in *Working Papers: Detroit*. March, 2004, edited by D. Mende and P. Oswalt. Available at: http://www.shrinkingcities.com [accessed: 1 July 2010].

Huffington Post. 2010. Detroit wants to save itself – By Shrinking. Available at: http://www.huffingtonpost.com [accessed: 23 August 2010].

Irwin, A. 1995. *Citizen Science*. London: Routledge.

Jack, I. 2010. Can manufacturing fill Britain's economic vacuum? *The Guardian*, 24 July.

Jacques, M. 1989. Britain and Europe, in *New Times – The Changing Face of Politics in the 1990s*, edited by S. Hall and M. Jacques. London: Lawrence and Wishart, 236–44.

Kay, K. 2010. Planting Detroit. *BBC News*, 5 August. Available at: http://newsvote.bbc.co.uk [accessed: 19 August 2010].

Kerr, J. 2000. Trouble in motor city, in *Working Papers: Detroit*. March, 2004, edited by D. Mende and P. Oswalt. Available at: http://www.shrinkingcities.com [accessed: 1 July 2010].

Lee-Boggs, G. 2009. Detroit's 'Quiet Revolution'. Available at: http://www.thenation.com [accessed: 16 July 2010].

Leonard, T. 2009. US cities may have to be bulldozed in order to survive. *The Daily Telegraph*. Available at: http://www.telegraph.co.uk [accessed: 23 August 2010].

Mason, P. 2010. Gary, Indiana: Unbroken spirit amid the ruins of the 20th Century. Available at: http://www.bbc.co.uk [accessed: 13 October 2010].

Mende, D. and Oswalt, P. 2004. Working Papers: Detroit. March, 2004. Available at: http://www.shrinkingcities.com [accessed: 1 July 2010].

Moceri, T. 2003. Devil's night, in *Working Papers: Detroit*. March, 2004, edited by D. Mende and P. Oswalt. Available at: http://www.shrinkingcities.com [accessed: 1 July 2010].

Moon, W. 2009. Reclaiming the Ruin. *The Design Observer Group*. Available at: http://places.designobserver.com [accessed: 23 August 2010].

Murray, R. 1989. Fordism and post-Fordism, in *New Times – The Changing Face of Politics in the 1990s*, edited by S. Hall and M. Jacques. London: Lawrence and Wishart, 38–53.

New Economy Task Force. 2000. *Making the New Economy Grow*. Washington, DC: Progressive Policy Institute. Available at: http://www.ppionline.org [accessed: 15 March 2007].

Orwell, G. 1950. *Nineteen Eighty-Four*. London: Signet.

Park, K. 2004. Shrinking city Detroit, in *Working Papers: Detroit*. March, 2004, edited by D. Mende and P. Oswalt. Available at: http://www.shrinkingcities.com [accessed: 1 July 2010].

Renn, A.M. 2009. *Detroit: Urban Laboratory and the New American Frontier*. Available at: http://www.newgeography.com [accessed: 23 August 2010].

Simon, B. 2010. Chinese making inroads into Detroit. *Financial Times*, 6 December.

Solnit, R. 2007. Detroit arcadia: Exploring the post-American landscape. *Harpers*, July. Available at: www.harpers.org [accessed: 25 August 2010].

Stoesz, D. 1992. The fall of the industrial city: The Reagan legacy for urban policy. *Journal of Sociology and Social Welfare*, 19, 149–67.

University of Michigan. 2010. Research Seminar in Quantitative Economics: Some highlights from the most recent RSQE Michigan forecast, released on May 27, 2010. Ann Arbor, Michigan: University of Michigan. Available at: www.rsqe.econ.lsa.umich.edu [accessed: 27 August 2010].

Urry, J. 1989. The end of organised capitalism, in *New Times – The Changing Face of Politics in the 1990s*, edited by S. Hall and M. Jacques. London: Lawrence and Wishart, 94–102.

Walsh, D. 2005. One-third of Detroit's population lives below poverty line. *World Socialist Web Site*. Available at: www.wsws.org/articles/2005 [accessed: 16 July 2010].

Whitford, D. 2009. Can farming save Detroit? *CNNMoney.com*. Available at: http://i2.cdn.turner.com/money/2009/12/29/news/economy/farming_detroit.fortune [accessed: 19 August 2010].

Zimmerman, A.D. 1995. Toward a more democratic ethic of technological governance. *Science, Technology and Human Values*, 20(1), 86–107.

10

Conclusion

Simon Bennett

What are we? Are we modern, late-modern or post-modern? Or a mix of all three? Do we live in a Risk Society, or are we still more worried about putting bread on the table than saving the planet? Are we all destined to work in offices or will some of us still work in factories and in the fields? Will one third of us be rendered unemployed or unemployable by poor education, under-achievement or new technologies? Academics have invested much in trying to align us with their particular world-view.[1] There is reputation (and money) to be made out of theorising.[2]

Detroit's experience suggests an uneven epochal messiness. In Detroit modernity co-exists with late-modernity in an unequal relationship: Detroit's motor and parts companies and large-scale urban farmers can summon significantly more economic and political capital than the city's innovative micro-farmers. Perhaps Steve Bruce (1999) is right. Perhaps epochs do overlap to produce a kaleidoscope of social, economic and political forms? Bruce puts it this way:

> *It is always useful to remind the intellectuals of London, Paris and New York that much of the life in the provinces goes on little changed .... The heavy industries of the Ruhr and the Clyde have disappeared, but workers are still organised in trade unions and occupational class still affects people's attitudes, beliefs and political behaviour ... the hard facts of people's lives, their health and longevity, remain heavily determined by class .... As we see when states limit trade, set quotas on*

---

1 The more actants you can align with (sign up to) your cause, the better your chance of winning out over the opposition.
2 I acknowledge that I, too, stand to gain from theorising that those who theorise post- and late-modernity may have got it wrong. I also acknowledge that in making this point I show a late-modern predilection for reflexivity.

immigration ... the announcement of the passing of the nation state is, to say the least, premature.[3]

David Cameron's Big Society idea evidences the tensions between modern and late-modern social formations. Yes, the Prime Minister wants to encourage grass-roots activity, and yes, the government will provide advice and financial support to active citizens[4] – but the State will not wither away. Indeed, all the indications are that the State will seek greater influence: during the financial crisis it extended its reach by investing in the financial sector (to prevent the collapse of the banking industry). There are centrifugal and centripetal forces at work. Today's watchwords are (or should be) concurrency and parallelism. The potential for tension and conflict is obvious (as in Detroit).

It is against this uncertain and fractious context that the well-intentioned seek to plan for and manage risk, crisis and disaster. It could be argued, of course, that as far as emergency planners and responders are concerned, the academic debate over consciousness and epochal shift is just that – academic. What matters is whether a specific public responds to an appeal for volunteers, or whether a group of professionals is open-minded enough to capitalise on citizens' craft (non-systematic) knowledge.

Given the 'patchwork' character of our time, the only certainty is that one cannot be certain that a particular model will work. Context – and in this author's opinion, economic prosperity – colours our risk consciousness. Consider, for example, the United Kingdom in 2011. Is it not possible that economic worries (job-shedding in the public and private sectors, inflation and the possibility of a double-dip recession) will act to undermine concern for the environment or our fellow citizens?[5] What is more important to someone in marginal

---

3   In December 2011 David Cameron used Britain's veto to opt-out of a new economic crisis-inspired European Union Treaty (Grice 2011a). Could Cameron's popular (in Britain, at least) re-assertion of statehood mark the beginning of the end for the European Union's 60 year drive for 'ever-closer union'? O'Flynn (2011) observed of the EU's economic malaise: 'Were countries such as Italy, Greece, Spain and Portugal to fall out of the Euro and bring back their own currencies, then the whole ideology of ever-closer union would have failed'. Much to the irritation of its EU partners, Britain has retained its border controls. By the end of 2011 Anglo-French relations had reached a new low with ministers from both governments trading public insults (Willsher 2011).

4   Some argue that Cameron's Big Society concept is no more than a cover for the most dramatic expenditure cuts for decades.

5   Britons would not be alone if they re-ordered their priorities. In 2011 Canada's right-wing government effectively withdrew from the Kyoto Treaty on greenhouse gas emissions, allowing it to exploit Canada's tar sands (a 'dirty' source of oil) unencumbered. According to Usborne (2011) 'A new survey published by the Institute for Research and Public Policy found that nearly two thirds of Canadians think their country is on the "right track"'. Circumstance

employment – working as hard as you can in the hope that you will keep your job, or volunteering to run a youth project or library or wildlife sanctuary? Beck would have us believe that we no longer have to struggle for our daily bread. Try persuading the marginally employed of that. Try telling a single parent living on a sink estate in Birmingham that recycling their rubbish or protesting against nuclear power or a local incinerator plant is more important than keeping your unemployed and disenchanted son or daughter off drugs. People only have so much physical and emotional energy.[6] The world looks different when you are in the gutter, or just off it. The case against (glib) generalisation is eloquently made in A Seventh Man:

> *To try to understand the experience of another it is necessary to dismantle the world as seen from one's own place within it, and to re-assemble it as seen from his. For example, to understand a given choice another makes, one must face in imagination the lack of choices which may confront and deny him .... The world has to be dismantled and re-assembled in order to grasp, however clumsily, the experience of another ... The subjectivity of another does not simply constitute a different interior attitude to the same exterior facts. The constellation of facts of which he is the centre is different (my emphasis). (Berger and Mohr 1975)*

However one labels the current epoch, one thing is certain – these are precarious times. Not as precarious as living through a world war, perhaps, but precarious none the less. Workers face numerous risks. Even the well-educated 'top third' face risks. Reflecting on his time as a high-earning contractor Smallman

---

colours perception: 'The Conservatives ... know that although most Canadians like to think of themselves as green, they are more worried just now about the health of the economy than that of the planet' (*The Economist* 2011). In 2002 Canada's future Prime Minister described the Kyoto Treaty as a 'job-killing, economy-destroying pact ... a socialist scheme to suck money out of wealth-producing nations' (Harper cited in *The Economist* 2011). The point has been made that the most polluting nations emit the most pollution because they produce the bulk of the world's goods and services. What proportion of the world's goods and services are produced by, say, the Maldives? Each year the global population grows by circa 80 million. In 2011 the global population passed seven billion. How is the world's rapidly growing population to be fed, clothed, housed and gainfully employed without increased industrial activity? How are the United Nation's Millennium Development Goals (2011) to be achieved without increased industrial activity?

6 In March 2011, the BBC broadcast a series of documentaries featuring the English actor Neil Morrissey. In the series Morrissey investigated why he had spent so much of his childhood in care. It transpired that his parents were so exhausted by their shift-work that they could not prevent the Morrissey household from descending into chaos. The more chaotic it became, the less time Neil's father spent at home. A downward spiral of neglect ensued. Lacking boundaries, Neil and his brother took to stealing. They spent nights alone at home. Eventually Social Services intervened. The brothers were sent to separate care homes. The experience coloured the rest of Neil's life.

observes: 'Was there a dual labour market? Without doubt, but it exhibited both sides of the dialectic. [I] was a high earning contractor, but with a role that could be terminated at any time'. Smallman and Robinson see in Chicago's Wacker Drive a metaphor for contemporary social and economic stratification. The rich get richer and the poor sink (on Wacker drive, literally, to a vile undercroft).[7]

As Bennett notes, even pilots' terms and conditions are under pressure. Once a glamorous occupation, commercial aviation is being de-professionalised and de-skilled. In a volatile industry pilots are obliged to 'follow the work'. This dynamic is creating a 'pilot diaspora' – an itinerant army of pilots (and engineers and cabin crew) who take up residence in high-density, sub-standard accommodation in order to keep their jobs. In the Risk Society there is a gulf between image and experience. In the public's imagination pilots (and cabin crew) enjoy a glamorous lifestyle. They fly to exotic destinations, occupy the best hotels, spend their days by the pool. In reality their salaries are under pressure, they are worked as hard as regulations will allow and are redeployed with little regard to personal circumstance. As the 2009 Colgan Air accident demonstrated, lifestyle reconfigurations have social, economic and safety implications. Rebecca Shaw commuted right across the USA to get to work. Marvin Renslow commuted from Florida. Neither had accommodation at Newark International. The accident highlighted the potential safety impacts of long-distance commuting and the increasing sordidness of the pilot lifestyle. In the Risk Society corporations call the shots, seemingly inured to collateral damage (Bauman 1998). The imperative seems to be to maximise the return on investment (while always acting within the law).

In the Risk Society the personal is sacrificed on the altar of profit. In the world of work choice is circumscribed. The age in which we live is an age of contradictions. Yes, we have more TV channels to choose from, and yes, we can holiday in myriad exotic locales, and yes, we can express our sexuality without fear of victimisation (generally), and yes, we can dress more or less

---

7 Across the globe the gulf between rich and poor continues to widen (International Labour Office 2011). A 2011 survey of British voters revealed that 70% considered the gap between rich and poor to be too wide (Grice 2011b). According to the Resolution Foundation, economic inequality affects life-chances. Garner (2011) explains: 'Research ... shows that even children from low-to-middle income families ... started school five months behind their richer peers in their skills with vocabulary, and were more poorly behaved'. Food banks, an established feature of North American charity efforts, are increasingly common in the United Kingdom (Insley 2011). One charity worker observed of those who used food banks at Christmas 2011: 'We were expecting to see an increase, but have been overwhelmed by the level of demand .... They are families who have hit hard times, people who have been made redundant, become ill or who are experiencing benefit delay' (Quinn cited in Keates 2011).

as we like, and yes, we can use surgery to re-engineer our body-shape, but no, we can no longer expect a job for life. Surely the most pertinent question is: Is choice (often no more than a matter of choosing which frivolity one samples) an adequate substitute for security?[8]

According to Alford, information and communication technologies are Janus-faced. On the one hand they can facilitate personal and societal liberation – witness the role of communications technologies in the Egyptian and Libyan revolutions, for example.[9] On the other, they can de-skill and de-personalise. Lacking the means to evaluate the consequences of rapid technological development, publics have little choice but to 'go with the flow'. In informal settings (the home, for example) ICTs can enrich and liberate. In formal settings (work, for example) they can dehumanise and imprison. The solution, suggests Alford, may lie in a broader public participation in technology assessment.

Both Miles and Masys argue for the recognition and utilisation of multiple realities within broadly similar domains – the emergency services (the 'blue light' services) and corporate responses to major incidents and accidents. Complexity produces risk and opportunity. The possibility of unforeseen interactions within complex socio-technical systems generates risk (Perrow 1984, 1999). As Lagadec (1993) puts it:

---

8   Regarding the Chinese experience, one might ask whether consumer choice is an adequate substitute for freedom of thought and action? China's unelected rulers may have embraced consumerism, but the country remains under the heel of a profoundly undemocratic and intimidatory dictatorship. Consumerism is a banal freedom.

9   The Mubarak regime tried, with little success, to disrupt communications. The Gaddafi regime also tried to disrupt communications during the civil unrest that began in late February, 2011. Both regimes fell to popular uprisings. Suspicions that Russia's December 2011 parliamentary elections had been stolen by the Putin-Medvedev axis led to protests in several Russian cities. The protests were underwritten by social media reporting. Outmaneuvered, Russia's state-controlled terrestrial channels had no choice but to cover the rallies. According to Brown (2011): '[I]t appears that someone high up in the Kremlin had realised that with so much discussion of the rallies online, the television would look ridiculous if it did not cover it'. *The Economist* (2011) offered this analysis of the demonstrations (the largest of which summoned around 100,000 onto the streets): '[Protesters] were mobilised by social networks rather than political parties. A key figure was Alexei Navalny, a popular blogger [web logger] who has used social networking to shake the Kremlin's power and undermine United Russia [Putin's party of nationalists] …. He circumvented the state monopoly on television by appealing to an internet audience'. Internet activism resonates with Russia's growing middle class who are urbane, travelled, well-educated, insulated from downturn and no longer willing to accept the old political settlement (under which citizens were allowed to enrich themselves so long as they turned a blind eye to corruption in government, business and policing (the same dynamic obtains in China), and to the associated maneuverings of Medvedev and Putin). Christmas Eve 2011 saw the largest demonstration, when circa 100,000 Muscovites took to the streets. They called for an end to Putin and Medvedev's 'managed democracy' (as a method of governance surely a paradox?) (Shuster 2011).

> *Major crises – from Challenger, Bhopal, Tylenol or Chernobyl to Exxon*
> *Valdez and Braer – are no longer exceptional events. Indeed the risk*
> *of crisis is even becoming structural as large networks become more*
> *complex, more vulnerable and more independent …*

Paradoxically, complexity also creates new opportunities for managing risk (Browning and Shetler 1992). The multi-vocality of late-modernity potentially provides new ways of seeing and doing. Only potentially, because, as Bruce (1999) explains, the old power-blocs and vested interests remain.[10] Why, for example, should doctors risk undermining their status by recognising lay knowledge? Why should emergency service chiefs risk their image and status by acknowledging fire officers' or paramedics' insights? Does human nature reflect epochal shifts, or remain constant?

More fundamentally, perhaps, can multi-vocality find expression in a world still dominated by powerful elites? As Alford points out, there are many barriers to public participation in technology assessment. Obstacles range from a disinterested and potentially obstructive technocratic elite to citizens' (perceived?) lack of time and expertise. As mentioned above, there is every possibility that, over time, Detroit's grass-roots urban farming movement will be absorbed by agribusiness. In Britain the Prime Minister's flagship Big Society initiative has started slowly amid a chorus of criticism and complaint. David Cameron relaunched the initiative in February 2011 amid accusations that it was nothing more than a cynical attempt to mitigate the effects of significant cuts in central and local government spending. *The Guardian* (2011) attributed the slow start to the public's (perhaps understandable) preoccupation with prices, incomes and jobs:

> *[A]sking individuals to do a bit more and the state to do a bit less is a*
> *philosophy for optimistic and prosperous times. It makes far less sense*
> *when times are insecure and people have such pressing anxieties and*
> *needs.[11] These are times when the state, though constrained and short*
> *of resources, is needed more than ever by many.*

---

10 Most of the world's wealth is controlled by a very small number of people. It is this concentration of wealth that inspired the Occupy movement of 2011. A 2006 study by the United Nations University revealed that: 'The richest 2 per cent of people in the world own more than half of all household wealth, while the poorer half of the global population control just 1 per cent' (United Nations News Centre 2006).

11 At the close of 2011 the British public faced rising unemployment, cuts in social security benefits and, in the event of the collapse of the Eurozone, the risk of a slump more severe than that seen in the 1930s.

Could it be that a society's tendency to multi-vocality, voluntarism and participation is partly a function of how secure that society feels? Could it be that the tougher the going, the more insular and self-interested the public becomes? Was Thatcher right? Is there no such thing as society?

If social conscience is so closely tied to feelings of security, where does that leave the environmental movement? When Beck first wrote about environmental consciousness, living standards were rising. Now, in many European countries, they are either stagnating or falling – according to Hari (2011) a consequence of our subsidising bankers' and shareholders' greed and stupidity.[12] While there is some evidence to suggest that citizens' enthusiasm for green issues has waned,[13] the fact of environmental disamenity is indisputable (consider the near-disappearance of the Aral Sea, for example). The built environment is still threatened by natural and man-made hazards – as the citizens of north-eastern Japan discovered in March 2011. At the end of March 2011 radioactive matter ejected by Tokyo Electric's breached Fukushima plant was discovered in Scotland. Radiation is a transboundary risk par excellence.[14]

Risk is an enduring theme of modern life. It permeates the political, economic and environmental domains. Some risks (earthquakes and tsunamis, for example) are unavoidable. Others are not. As Waddington showed, a more holistic analysis of the causes of terrorism may help reduce the risk of terrorist attack. Acceptance of the limitations of free-market solutions to urban decay[15]

---

12 At the time of writing (2011) the spectre of stagflation (not seen since the darkest days of the 1970s) haunted the British economy.

13 A public opinion survey conducted by the US Pew Research Centre found relatively little concern about global warming: 'According to the survey of 1,503 adults, global warming, on its own, ranks last out of 20 surveyed issues. Here's the list from top to bottom, with the economy listed as a top priority by 85 percent of those polled and global warming 30 percent: the economy, jobs, terrorism, Social Security, education, energy, Medicare, health care, deficit reduction, health insurance, helping the poor, crime, moral decline, military, tax cuts, environment, immigration, lobbyists, trade policy, global warming' (Revkin 2009).

14 Cumbrian hill-farms are still contaminated with fallout from Chernobyl. The Cameron–Clegg government's Fukushima-inspired review of nuclear safety found no major problems with the UK nuclear industry: '[T]he review … concluded that the UK's nuclear industry is broadly safe, with "no fundamental safety weaknesses". If the areas of concern raised in the light of the Fukushima [event] are addressed, the industry will be "even safer", the report said. The relatively clean bill of health was rapidly seized on by the government' (Harvey, Vidal and Edwards 2011). The Cameron–Clegg government frames (constructs) nuclear power as a carbon-neutral (and therefore environmentally benign) source of baseload. Even politicians who used to be against nuclear power (like Energy and Climate Change secretary Chris Huhne) are now on board.

15 While the Thatcher government's London Docklands Development Corporation (LDDC) paved the way for London's new financial centre, the development catered more for a highly educated and handsomely rewarded transnational workforce than for the indigenous populations

and recognition of the potential benefits of spontaneous civic action may save the inner cities. Official, top-down initiatives to deal with disaffection amongst urban youth and disintegrating inner-city economies have met with mixed success. The 2011 riots which saw widespread public disorder, looting and arson in several cities, including London and Birmingham (England's first and second cities) were due in part to feelings of antipathy towards the police: 'Widespread anger and frustration at the way police engage with communities was a significant cause of the summer riots in every major city where disorder took place, the biggest study into their cause has found. Hundreds of interviews with people who took part in the disturbances which spread across England in August revealed deep-seated and sometimes visceral antipathy towards police .... Rioters revealed that a complex mix of grievances brought them on to the streets but analysts appointed by the [London School of Economics] identified distrust and antipathy toward police as a key driving force .... Although rioters expressed a mix of opinions about the disorder, many of those involved said they felt like they were participating in explicitly anti-police riots' (Lewis, Newburn, Taylor and Ball 2011). Reactions to the worst public disorder for a generation ranged from calls for more investment in inner-city economies, the restoration of financial support for students (young people have suffered more than others in the post-2008 downturn (Moore 2011))[16] and the curtailment of stop-and-search sweeps, to allowing police to use water cannon, tear gas, Tazers, rubber bullets and live ammunition against rioters (Sheldrick and Twomey 2011). Britain's authoritarian right-wing tabloids (the *Daily Mail*, *Daily Star*, *Daily Express* and *The Sun*) made much of calls for a more vigorous response to disorder. The front page of the December 21 2011 edition of the *Express* carried the (celebratory?) banner: 'Now Police Can Shoot Rioters'.[17]

---

of Tower Hamlets, Newham, Hackney and other inner-city London boroughs. The 1980s protestations of local MPs, councillors, activists and communities went more or less unheeded by the LDDC. Today locals are far more likely to be employed as low-paid casualised office cleaners than as commodity traders. Canary Wharf reproduces inequality. Its towers reify the power of unanchored (and largely unaccountable) transnational capital (Bennett 2009).

16 According to Moore (2011): 'The economic cost of the misdeeds of their parents has fallen squarely on the young .... The trouble is that if you squeeze something too hard it has a tendency to pop'. England's youth popped in August 2011.

17 A 2011 report from Her Majesty's Inspectorate of Constabulary (HMIC) suggested that firearms could be used in riot situations when human life was in jeopardy (perhaps because of mob violence or arson attacks) (Sheldrick and Twomey 2011). The police took small arms fire during the 2011 riots. A gun was trained on a police helicopter. Targeted interventions in inner-city communities can produce results. Investments in Manchester's Moss Side neighbourhood have helped reduce youth crime: 'In the past decade Moss Side has started to change thanks to a range of youth services .... They have worked hard to change the decades of despondency that once blighted the area .... It has meant a real result – the steady decline of crime' (Webb 2011). Unfortunately, many of these services now face cutbacks or dissolution. Some claim the Big Society will fill the void.

Applying HRO theory to the emergency services may improve outcomes. Working with, rather than against nature, may reduce flood risk. Framing global warming more as an opportunity than a burden may help us find more effective solutions. Involving the public in technology risk assessment may achieve a more benign application of new technologies. Recognising the contingency of organisational culture as it pertains to safety may help save lives, hardware and reputations. Subsidising accommodation for low-paid flight crew may improve aviation safety. Theorists like Beck and Perrow paint a bleak picture. Their world is a world of multiplying transnational risks, runaway technologies and normal accidents. However, as this book demonstrates, the problems we face in the twenty-first century are not intractable. All it takes is a little less dogma and a little more imagination. We are too much prisoners of the past.

References

Bauman, Z. 1998. *Globalisation: The Human Consequences*. New York: Columbia University Press.

Bennett, S.A. 2009. *Londonland: An Ethnography of Labour in a World City*. Faringdon: Libri Publishing.

Berger, J. and Mohr, J. 1975. *A Seventh Man*. London: Penguin.

Brown, J. 2011. No longer the same old story. *i newspaper*, 14 December.

Browning, L.D. and Shetler, J.C. 1992. Communication in Crisis, Communication in Recovery: A Postmodern Commentary on the Exxon Valdez Disaster. *International Journal of Mass Emergencies and Disasters*, 10(3), 477–98.

Bruce, S. 1999. *Sociology: A Very Short Introduction*. Oxford: Oxford University Press.

Dyson, J. 2010. *Ingenious Britain. Making the UK the leading high-tech exporter in Europe. A report by James Dyson, March, 2010*. London: Conservative Party UK.

Garner, R. 2011. Less affluent children are left behind. *i newspaper*, 14 December.

Grice, A. 2011a. Huhne: Tory right wants UK to be semi-detached member of EU. *The Independent*, 23 December.

Grice, A. 2011b. Public call for business to put people before profit. *i newspaper*, 24 December.

Hari, J. 2011. When will the soufflé of spin collapse? *The Independent*, 11 February.

Harvey, F., Vidal, J. and Edwards, R. 2011. UK nuclear safety review finds 38 cases for improvement. *The Guardian*, 11 October.

Insley, J. 2011. Plight of the families who must rely on food parcels. *The Observer*, 18 December.

International Labour Office. 2011. *Global employment trends 2011: The challenge of a jobs recovery*. Geneva: International Labour Office.

Keates, H. 2011. Service helps feed 'hidden hungry' in city. *South Wales Evening Post*, 23 December.

Lagadec, P. 1993. Ounce of Prevention Worth a Pound in Cure, *Management Consultancy*, June.

Lewis, P., Newburn, T., Taylor, M. and Ball, J. 2011. Rioters say anger with police fuelled summer unrest. *The Guardian*, 5 December.

Moore, J. 2011. Outlook. *The Independent*, 23 December.

O'Flynn, P. 2011. Britain's future lies out of the EU and into the world. *Daily Express*, 17 December.

Perrow, C. 1984. *Normal Accidents: Living with High-Risk Technologies*. New York: Basic Books, Inc.

Perrow, C. 1999. *Normal Accidents: Living with High-Risk Technologies* (2nd edition). Princeton, NJ: Princeton University Press.

Revkin, A.C. 2009. Obama Urgent on Warming, Public Cool. *The New York Times*, 22 January.

Sheldrick, G. and Twomey, J. 2011. Now Police Can Shoot Rioters. *Daily Express*, 21 December.

Shuster, S. 2011. Putin on the back foot after 100,000 demand reform. *i newspaper*, 26 December.

The Guardian. 2011. Big society: Good ideas, bad time. The time for the big society will come. But that time is not now, not in this way. *The Guardian*, 15 February.

The Economist. 2011. Kyoto and out. *The Economist*, 17 December.

The Economist. 2011. The birth of Russian citizenry. *The Economist*, 17 December.

United Nations News Centre 2006. Small band of rich controls majority of global wealth, UN study finds. New York: United Nations.

Usborne, D. 2011. PM pulls plug on Kyoto, stoking claims country is charging towards the Right. *i newspaper*, 14 December.

Webb, B. 2011. The Big Society will struggle in Moss Side. *The Guardian*, 14 February.

Willsher, K. 2011. Entente discordiale? Oh no, the French don't really hate us ... or do they? *The Observer*, 18 December.

World Health Organisation. 2011. *Millennium Development Goals: Progress Towards the Health-related Millennium Development Goals*. Available at: http://www.who.int/mediacentre/factsheets/fs290/en/index.html [accessed: 1 December 2011].

Index

Page numbers in *italics* refer to figures and tables.

If you have found this book useful you may be interested in other titles from Gower

Making Ecopreneurs
Developing Sustainable Entrepreneurship
Edited by Michael Schaper
Hardback: 978-0-566-08875-9
e-book: 978-1-4094-0123-0

Innovation and Marketing in the Video Game Industry
Avoiding the Performance Trap
David Wesley and Gloria Barczak
Hardback: 978-0-566-09167-4
e-book: 978-0-566-09168-1

Female Immigrant Entrepreneurs
The Economic and Social Impact of a Global Phenomenon
Edited by Daphne Halkias, Paul Thurman, Nicholas Harkiolakis
and Sylva Caracatsanis
Hardback: 978-0-566-08913-8
e-book: 978-0-566-08914-5

Creating and Re-Creating Corporate Entrepreneurial Culture
Alzira Salama
Hardback: 978-0-566-09194-0
e-book: 978-0-566-09195-7

Visit **www.gowerpublishing.com** and

- search the entire catalogue of Gower books in print
- order titles online at 10% discount
- take advantage of special offers
- sign up for our monthly e-mail update service
- download free sample chapters from all recent titles
- download or order our catalogue